Gravitation II –
relativistic planet orbit

U. KIVI

Gravitation II – relativistic planet orbit

Layout and Cover design: Books on Demand
Publisher: BoD – Books on Demand, Helsinki, Suomi
Manufacturer: BoD – Books on Demand, Norderstedt, Saksa

ISBN: 978-952-80-5699-7

CONTENT

FOREWORD

This book constitutes the second part for reference /1/: 'Gravitation, Exact calculation of Newton and Einstein theory'. In that book the gravitation equations are solved in an unusual way. The equations of Newton theory are constructed in a reverse order. Starting point is the observed elliptic orbit of a planet and the theory is derived by demanding that the planet acceleration points to the Sun, which is located at one of the ellipse focal points. In this way one gets in an easier way the planet position as a function of time and also the speed and acceleration of the planet. These results are used as the starting point for the relativistic calculations.

In reference /1/ the relativistic planet orbit Euler equations are constructed in a different way compared to the normal method and the equations also are different from the normal Euler equations. The solutions of these equations in case of free fall and in case of two dimensional planet orbit calculation are the same as the solutions of the normal Euler equations. The Euler equations are solved by inserting a trial function, which is a sum of the corresponding classic solution and an unknown function. In this way you get such an equation for the unknown function that can be solved. This is probably a new way to solve the planet orbit according to relativity theory. Using this method only one integrations constant appears into the solution function of the planet orbit. If you solve the Euler equations in the normal way you get two integrations constant into the solution function.

The results presented in reverence /1/ constitute the starting point for the theory presented in this book.

Helsinki 28.2.2021 Uuno Kivi

1. INTRODUCTION

The Euler equations for the planet orbit constructed in reference /1/ constitute the starting point for the theory presented in this book. A practical method for planet orbit calculation according to relativity theory is developed in this book. There appears two integration constants in the orbit equations constructed in the literature. It doesn't seem to be easy to determine prober values for these constants. If you construct the Euler equations in the way presented in reference /1/ you don't get any unknown integration constants in the equations, if you make the solution for the angular speed so that it approaches the classical solution in the classical limit (speed of light c → ∞). Mathematically there should be an unknown integration constant. In the equation for the angular speed you may multiply this constant into value for the half length of the orbit ellipse minor axis – b. In practice this constant had to taken in use in reference /1/, because you had to change the value for the minor axis to get the Sun to the same position in relativistic and classic solutions.

We proceed mathematically in a more accurate way in this book compared to reference /1/. We demand that the distances of the planet in perihelion and aphelion are the same according to relativity theory and according to classical mechanics and in this way we get equations for the integration constants as functions of perihelion and aphelion distances and Schwarzschild radius. One could also verify in this way that the correction for the half minor axis used in reference /1/ was quite accurate. This correction could be made in reference /1/, because the values for the major and minor axes are pure parameters in the relativistic solution without any geometric interpretation and the relativistic solution is not an ellipse although it is nearly an ellipse in a weak gravitation field. The minor axis half was increased in reference /1/ by 3018.779 m and the value of the major axis was not changed. In this way the Sun could be positioned with an accuracy of 0.4 mm into the same point in relativistic and classical solution. According to the more accurate solution presented in this book the minor axis half is increased by 3018.79877 m and the major axis half is increased by 0.00016 m. In this way the planet distances at perihelion and aphelion are exactly the same in relativistic and classical solutions.

2. PLANET ORBIT

2.1 Basic theory

Euler equations in polar coordinates may be written according to reference /1/ as (eq. 2.8.48 and 2.8.49).

(2.1.1)
$$\ddot{\varphi} = -\frac{2\dot{\varphi}\dot{r}}{r} + \frac{2\dot{\varphi}\dot{r}}{r}\frac{\alpha}{2(r-\alpha)}$$

(2.1.2)
$$\ddot{r} = r\dot{\varphi}^2 - \frac{MG}{r^2} - \alpha\dot{\varphi}^2 + \frac{\alpha MG}{r^3} + \frac{3\alpha\dot{r}^2}{2r(r-\alpha)}$$

where r is the planet distance from the Sun, φ is the planet polar angle coordinate, α is Schwarzschild radius, M is the mass of the Sun and G is the general gravitation constant.

By integrating equation 2.1.1 with respect to time t one gets (reference /1/ equation 2.8.61)

(2.1.3)
$$\dot{\varphi} = \sqrt{\frac{GM\bar{b}^2}{\bar{a}}\frac{r-\alpha}{r^3}}$$

Parameters $\bar{a}$ and $\bar{b}$ are otherwise arbitrary, but in the classical limit (c → ∞, a → 0) they have to approach major and minor axis halves of the classical orbit ellipse – $\bar{a}$ → a, $\bar{b}$ → b. Now equation 2.1.3 approaches the corresponding classical mechanics equation. If you insert expression 2.1.3 into equation 2.1.2 and solve $\dot{r}^2$ you get (reference /1/ expression 2.8.69)

(2.1.4)
$$\dot{r}^2 = \frac{MG(r-\alpha)^2}{r^3}\frac{2\bar{a}r^2-(\bar{b}^2+r^2)(r-\alpha)}{\bar{a}r^2}$$

You may eliminate time t from equations 2.1.3 and 2.1.4 and the result is relativistic orbit equation in polar coordinates (reference /1/ equation 2.8.71)

(2.1.5)
$$r_\varphi^2 == -\frac{r^4}{\bar{b}^2} + \frac{2\bar{a}+\alpha}{\bar{b}^2}r^3 - r^2 + \alpha r$$

This is a separable differential equation whose solution is (reference /1/ equation 2.8.72)

(2.1.6)
$$\int_{r_0}^{r} \frac{\bar{b}}{\sqrt{r(2\bar{a}r^2 - (\bar{b}^2 + r^2)(r - \alpha))}} dr = \varphi - \varphi_0$$

If you take the classical limit ($c \to \infty$, $a \to 0$, $\bar{a} \to a$, $\bar{b} \to b$) you get the corresponding classical mechanics equation (reference /1/ equation 2.8.73)

(2.1.7)
$$\int_{r_0}^{r} \frac{b}{\sqrt{r^2(2ar - (b^2 + r^2))}} dr = \varphi - \varphi_0$$

Symbolize planet smallest distance from the Sun with r_1 and the largest distance from the Sun with r_2. You may calculate that the half of the major axis and the square of the half of the minor axis may be calculated using equations

(2.1.8)
$$a = \frac{r_1 + r_2}{2}, \quad b^2 = r_1 r_2$$

Now equation 2.1.7 may be written as

(2.1.9)
$$\int_{r_0}^{r} \sqrt{\frac{r_1 r_2}{-r^2(r - r_1)(r - r_2)}} dr = \varphi - \varphi_0$$

The zero points of the 4-degree polynomial in the square root in the denominator of expression 2.1.7 are $r_{(0)} = 0$, $r_{(3)} = 0$, $r_{(1)} = r_1$ and $r_{(2)} = r_2$. The 4-degree polynomial is positive when $r_1 < r < r_2$ and expression 2.1.7 is real. Planet distance from the Sun in its elliptic orbit varies between the very same values $r_1 < r < r_2$.

One may assume that corresponding to the classical mechanics the distance r in the relativistic orbit equation (equation 2.1.6) varies between such values that make the square root in the denominator real. There seems to be no other justification for this than the analogy to classical mechanics solution (expression 2.1.7) and the feeling that distance r probably takes all the available values that it is able to take. To be able to compare the classic and relativistic solutions easily, we demand that the denominator in equation 2.1.6 is real with the same values for r that make the denominator real in the classic solution so that the expression under the square root in expression 2.1.6 should have the same zero points as the expression under the square root in expression 2.1.7 ($r = r_1$ and $r = r_2$). Let's define variables x and y in the following way $\bar{a} = a + x$ and $\bar{b}^2 = b^2 + y$.

Demand that r_1 and r_2 are the zero points of the 4-degree polynomial under the square root in expression 2.1.6. So that

$$(2.1.10) \qquad 2(a+x)r^2 - (b^2 + y + r^2)(r - \alpha) = 0$$

or

$$(2.1.11) \qquad 2ar^2 - (b^2 + r^2)r + 2xr^2 - (b^2 + r^2)\alpha - yr + y\alpha$$

$$= 2xr^2 - (b^2 + r^2)\alpha - yr + y\alpha = 0$$

must be zero at points $r = r_1$ and $r = r_2$. The solution for x and y is

$$(2.1.12) \qquad y = \frac{2\alpha ab^2}{b^2 - 2\alpha a}$$

$$(2.1.13) \qquad x = \frac{y}{4a} - \frac{\alpha}{2} = \frac{\alpha^2 a}{b^2 - 2\alpha a}$$

and

$$(2.1.14) \qquad \bar{a} = a + x = a + \frac{\alpha^2 a}{b^2 - 2\alpha a} = \frac{r_1 + r_2}{2} \frac{(r_1 - \alpha)(r_2 - \alpha)}{(r_1 - \alpha)(r_2 - \alpha) - \alpha^2}$$

$$(2.1.15) \qquad \bar{b}^2 = b^2 + y = b^2 \frac{b^2}{b^2 - 2\alpha a} = r_1 r_2 \frac{r_1 r_2}{r_1 r_2 - \alpha(r_1 + r_2)}$$

With these values for parameters $\bar{a}$ and $\bar{b}$ distances from the Sun in perihelion and aphelion are the same in relativistic and classical orbit. If we take the classic limit ($c \to \infty$, $a \to 0$), parameters $\bar{a}$ and $\bar{b}$ approach the corresponding classic values (equation 2.1.8).

The third root of equation 2.1.10 may be got using the identity

$$(2.1.16) \quad 2\bar{a}r^2 - (\bar{b}^2 + r^2)(r - \alpha) \equiv r^3 - (r_1 + r_2 + r_3)r^2 + (r_1 r_2 + r_1 r_3 + r_2 r_3)r - r_1 r_2 r_3$$

and the solution is

$$(2.1.17) \qquad r_3 = \frac{r_1 r_2 \alpha}{(r_1 - \alpha)(r_2 - \alpha) - \alpha^2}$$

You may now write the orbit equation as

$$(2.1.18) \qquad \int_{r_0}^{r} \sqrt{\frac{r_1 r_2 \frac{r_1 r_2}{r_1 r_2 - \alpha(r_1+r_2)}}{r(r-r_1)(r_2-r)(r-\frac{r_1 r_2 \alpha}{(r_1-\alpha)(r_2-\alpha)-\alpha^2})}} \, dr = \varphi - \varphi_0$$

or

$$(2.1.19) \qquad \int_{r_0}^{r} \frac{r_1 r_2}{\sqrt{r(r-r_1)(r_2-r)(r(r_1 r_2 - \alpha(r_1+r_2))-r_1 r_2 \alpha)}} \, dr = \varphi - \varphi_0$$

The orbit equation in expression 2.1.19 is given as a function of the known quantities r_1, r_2 and α that is as a function of the perihelion distance, aphelion distance and Schwarzschild radius.

2.2 Calculation methods

Using Euler equations as the starting point we have derived an equation for the planet orbit (equation 2.1.6) in polar coordinates r and φ. There appears two integration constants $\bar{a}$ and $\bar{b}$ in the expression. When values for the constants $\bar{a}$ and $\bar{b}$ have been chosen, solution for the planet orbit equation r = r(φ) is got by numerical integration of expression 2.1.6. To perform the numerical integration you have to know the integration limits for the distance r. In the classic orbit equation 2.1.7 r varies through all those values that make the square root in the denominator real ($r_1 < r < r_2$). Without eminent information we may assume that also in expression 2.1.6 r gets all those values which make the square root in the denominator real. So we may use the following method to calculate the relativistic orbit equation (method 1)

Method 1 to solve the orbit equation

1. Choose mass for the Sun or the black hole and calculate Schwarzschild radius α
2. Choose values for parameters $\bar{a}$ and $\bar{b}$.
3. Integrate expression 2.1.6 numerically from value r = r_1 to value r = r_2, where r_1 and r_2 are the largest zero points of the 4-degree polynomial under the square root. The 4-degree polynomial is positive, when $r_1 < r < r_2$

4. Now you have calculated half a round. If you want to calculate further the orbit, you may integrate numerically expression 2.1.6 from r_2 to r_1.
5. If you want to calculate more rounds, repeat parts 3 and 4.

This is a natural way to solve the planet orbit, you have an equation for the orbit and you solve it. To make the solution you need to know the roots of the 4-degree polynomial f(r). In the FORTRAN program presented in appendix 1 the roots are determined using two different methods (appendix 3) – subprograms NUMROOTS (numeric solution) and ANAROOTS (analytic solution).

Method 2 to solve the orbit equation

If you know the distance of the planet at perihelion and aphelion, it is natural to use method 2 to solve the orbit equation. Now parameters r_1, r_2 and α are given and parameters $\bar{a}$ and $\bar{b}$ are calculated using expressions 2.1.14 and 2.1.15. This method is preferable if you want to compare classic and relativistic orbit solutions for a certain planet. In this case the method to solve the orbit equation is as follows

1. Choose mass for the Sun or for the black hole and calculate Schwarzschild radius α
2. Choose values for the planet distance from the Sun at perihelion and aphelion – r_1 and r_2
3. Calculate values for parameters $\bar{a}$ and $\bar{b}$ using expressions 2.1.14 and 2.1.15.
4. Integrate expression 2.1.6 numerically from r_1 to r_2.
5. Now you have calculated half a round. If you want to calculate further the orbit, you may integrate numerically expression 2.1.6 from r_2 to r_1.
6. If you want to calculate more rounds, repeat parts 4 and 5.

2.3   FORTRAN-program to solve the planet orbit

The relativistic planet orbit is solved by numerically integrating expression 2.1.6. The classic solution for planet orbit is got by numerically integrating expression 2.1.7. The calculation may be performed using either of the two methods described above (variable MOD either -1 or 1). Listing of the program that performs the numerical integration is presented in appendix 1. Input is very short so it is written into the program. The program version listed in appendix 1 has been used to calculate calculation case 1 that is the calculation of Mercury orbit according to relativity and according to classical mechanics. The result of calculation case 1 is presented in chapter 3.1. Calculation case 1 is calculated using calculation method 2, so that it is easy to compare the classic solution with the relativistic solution.

In calculation cases 2 – 8 we examine how the variation the mass of the Sun (or black hole) affects the planet orbit. In all these cases the calculation method 1 is used. In all cases parameters $\bar{a}$ and $\bar{b}$ have been given the classic values of Mercury orbit (half of the major axis and half of the minor axis) that is

(2.3.1) $$\bar{a} = a = 5.79091755E10$$

(2.3.2) $$\bar{b} = b = 5.66716379E10$$

The Sun mass has been multiplied by 0.4E6, 1.0E6, 2.522090E6, 2.522091E6, 3.0E6, 30E6 and 300E6 in different cases and the corresponding Schwarzschild radius has been calculated. The planet distances at perihelion and aphelion have been got by solving the roots of the 4-degree polynomial

(2.3.3) $$f(r) = r\left(2\bar{a}r^2 - (\bar{b}^2 + r^2)(r - \alpha)\right) = 0.$$

The square root of polynomial f(r) is found in the denominator of expression 2.1.6.

The changes of the program input compared to calculation case 1 are described in connection with the calculation case in question.

2.4   Classic solution for planet orbit

We shall first examine the classic solution for the planet orbit to get a good reference for other calculations. We shall use the parameters of Mercury orbit: distance at perihelion $r_1 = 46001272000.0$ and distance at aphelion $r_2 = 69817079000.0$. Then you may calculate with the help of expression 2.1.8 values for the half major axis and for the half minor a = 57909175500, b = 56671637009.39376. We have used the following values for the constants: speed of light c = 2.99792458E8, mass of the Sun M = 1.9891E30, general gravitation constant G = 6.67428E-11. The expression under the square root in the denominator in equation 2.1.7 is a 4-degree polynomial

(2.4.1) $$f(r) = r^2(2ar - (b^2 + r^2))$$

and its graph is drawn in figure 2.4.1. The classic solution for planet orbit $r = r(\varphi)$ is got by integrating expression 2.1.7 numerically from value $r = r_1$ to value $r = r_2$ and then from value $r = r_2$ to value $r = r_1$ and so on. Here r_1 and r_2 are the largest roots of the 4-degree polynomial f(r). There exist also an analytic solution, which – as is well known – is an ellipse.

As is usual for a four degree polynomial, which opens downwards, the graph has two maxima and a minimum between the maxima. In this case the zero points of the polynomial are: double root $r_{3,4} = 0$ and $r_1 = 4.6001272E10$ and $r_2 = 6.981708E10$. The planet distance r varies between r_1 and r_2 in its elliptic orbit and the 4-degree polynomial is positive when $r_1 < r < r_2$ and the square root in expression 2.1.7 is real.

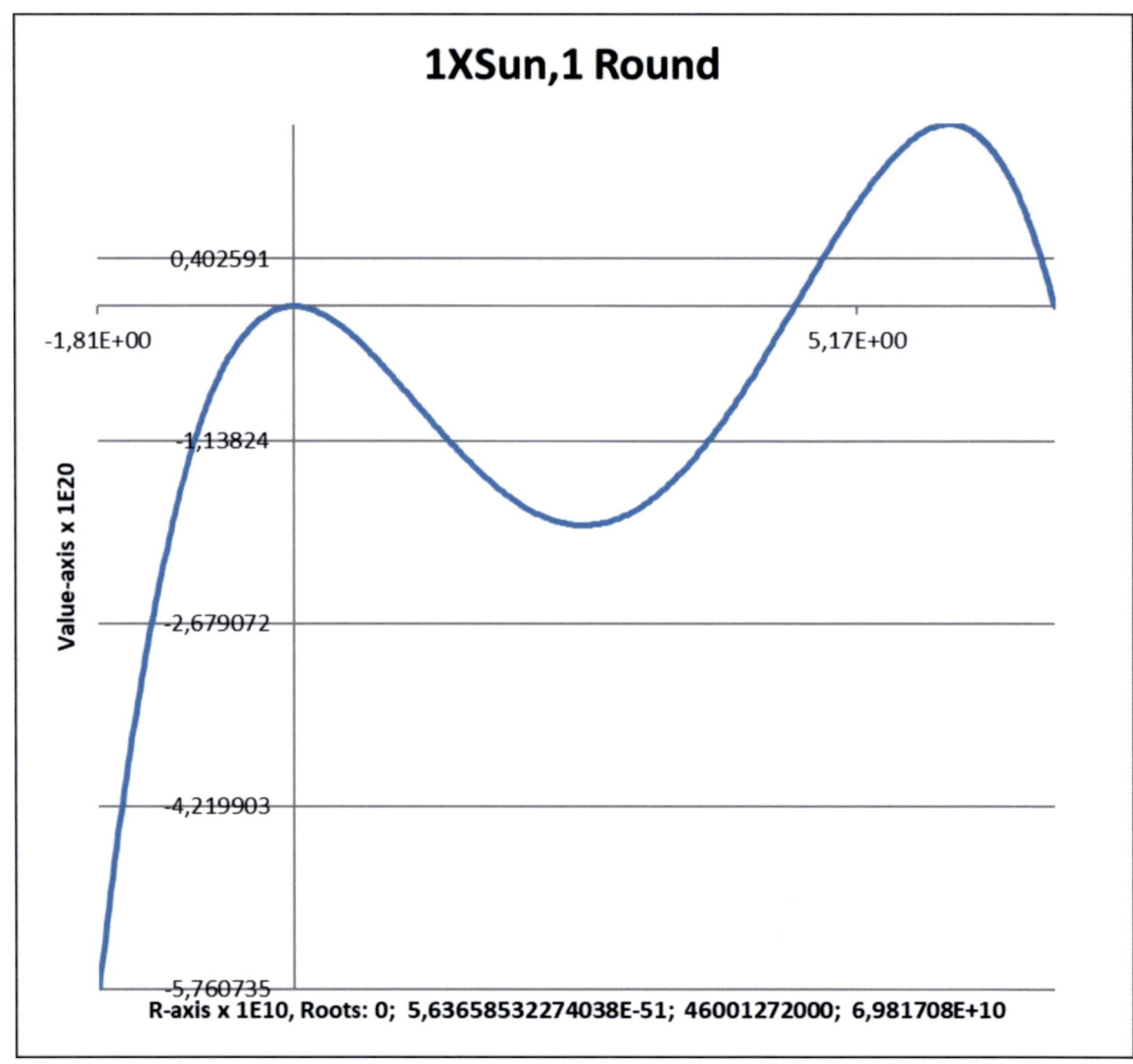

Figure 2.4.1: The 4-degree polynomial at classic mercury orbit

2.5   Relativistic planet orbit

Relativistic orbit for a planet may be calculated by integrating expression 2.1.6 numerically (to my knowledge there exists no analytic solution). Expression 2.1.6 differs from expression 2.1.7 only so that Schwarzschild radius α in the 4-degree polynomial under the square root has a nonzero value. Looking the expression for polynomial f(r) you see that, as α increases, the value of f(r) increases with positive r and the increase is the larger the larger value for r is considered. Also the zero points of polynomial f(r) change as α increases. Because the whole graph of f(r) increases with positive r, also the value of the minimum of f(r) increases first to value zero and then to positive values. You may deduce by looking expression 2.1.6, that a certain change Δr for the distance r causes the larger change in angle φ the smaller value f(r) is at that point. The closer to the r-axis the minimum value of polynomial f(r) lies the larger is the section for r at which f(r) is small near the minimum. Then angle φ increases very much even with small changes in r. In the extreme case f(r) has value 0 at the minimum (that is at perihelion $r = r_1$). In that case the planet goes around the black hole forever approaching infinitesimally the perihelion distance in every round trip (look calculation cases in chapters 3.4 and 3.5). It is physically sensible to think that planet distance uses all permitted values that is all the values that make polynomial f(r) positive. Then the integration in expression 2.1.6 has to be performed from value $r = r_1$ to value $r = r_2$, where r_1 and r_2 are the largest zero points of polynomial f(r).

The graph of the polynomial in case of Mercury orbit is drawn in figure 2.4.2. The graph in figure 2.4.2 differs only little from the graph in figure 2.4.1. In the calculation case 1 presented in chapter 3.1 we use calculation method 2 (orbit is calculated by numerical integration of expression 19) to determine the orbit, so the corresponding 4-degree polynomial graph is neither of the graphs in figures 2.4.1 or 2.4.2. This polynomial has the same two largest roots as the polynomial in figure 2.4.1, but the third root is almost the same as in figure 2.4.2. This graph does not differ with the resolution of the figures from the graphs in figures 2.4.1 and 2.4.2, so no own picture has been drawn for it.

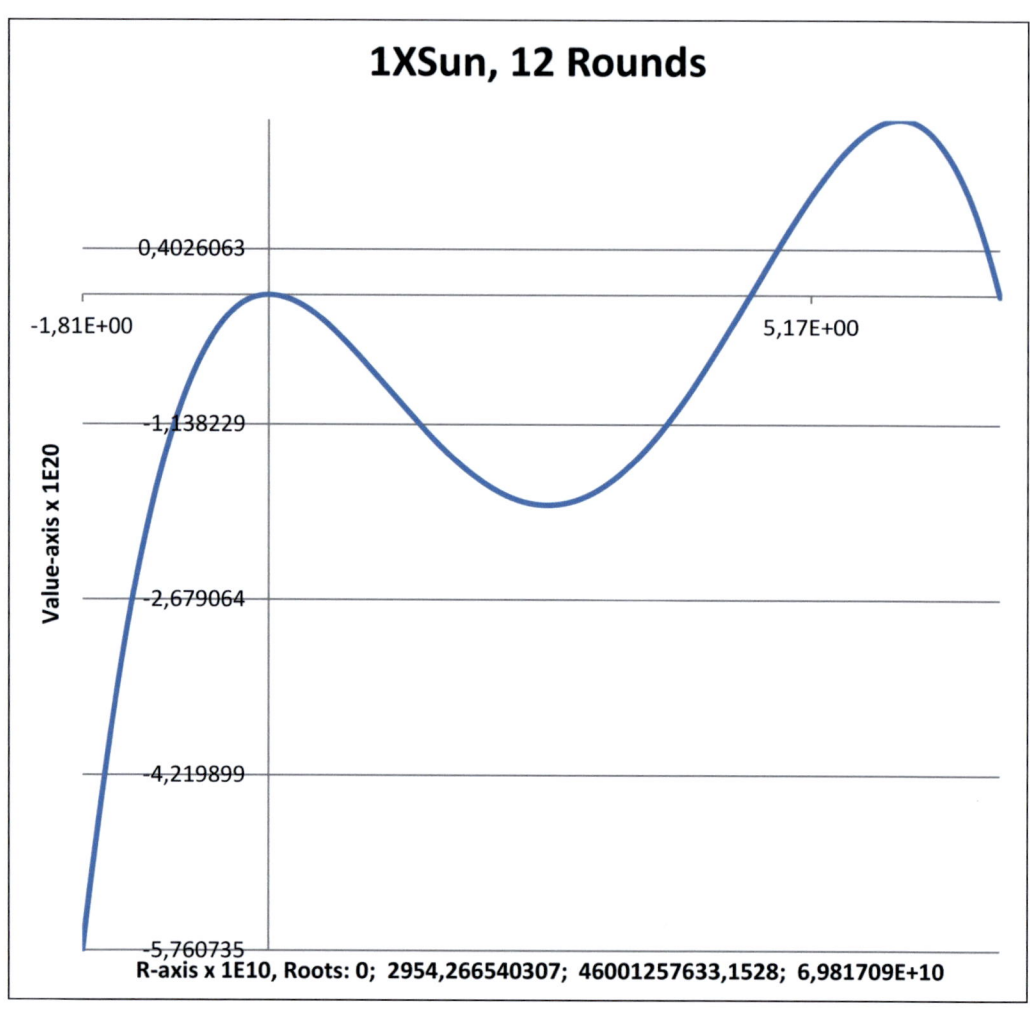

Figure 2.4.2: 4-degree polynomial at Mercury orbit.

3   CALCULATION CASES

3.1   Calculation case 1:
Mercury orbit according to Newton and Einstein

We shall use in the first calculation values for the parameters corresponding to the Mercury orbit: radius at perihelion r_1 = 46001272000.0 and radius at aphelion r_2 = 69817079000.0. We may then calculate values for the halves of the major axis and the minor axis a = 57909175500, b = 56671637009.39376 using expressions 2.1.8. Constants have been given values: speed of light c = 2.99792458E8, mass of Sun M = 1.9891E30, general gravitation constant G = 6.67428E-11. We may calculate then value 2954.27 for the Schwarzschild radius. We shall use calculation method 2 so that the orbit is solved by integration of expression 2.1.19 (or equivalently by integrating expression 2.1.7 with parameter values $\bar{a}$ and $\bar{b}$ calculated using expressions 2.1.14 and 2.1.15 that is with values $\bar{a}$ = 57909175500.00016 and $\bar{b}$ = 56671640028.17253). When one uses this method, the planet distances at aphelion and perihelion have the same values in classic and relativistic solutions. Because the value for distance r is monotonically decreasing, when planet proceeds from aphelion to perihelion, one may easily compare the relativistic and classic azimuth angles with the same value for r and get in this way local (differential) perihelion shift. We shall calculate 12 round trips around the Sun to examine the stability of the perihelion shift.

The relativistic and classic azimuth angles are calculated by numerically integrating expressions 2.1.6 and 2.1.7. The integration is performed for both expressions from value r_2 to value r_1. There are 7 basic intervals and their relative lengths are 1, 1E4, 1E7, 1E12, 1E7, 1E4, 1. Intervals have been divided into subintervals in the following way: 5E7, 1E8, 1E8, 1E8, 5E7, 1E8, 5E7. This is the basic division for the numerical integration. This division can be divided with a common divider. Values 1 or 5 have been used for the common divider in every calculation of this book. The division is actually too rough and a value such as 5000 should be used. The rough division doesn't seem to affect much the form of the orbit or the value of the perihelion shift calculated using equation 3.1.1. The error is visible, when comparing the distance r to the distance of the classical orbit ellipse calculated using equation 3.1.3. The rough division is necessary, because the calculation would take too long a time, when calculating several rounds around the

Sun with a dense division. The calculation of half a round takes many weeks with the best laptop that I own, if a dense nodal division is used.

The perihelion shift (arc seconds in 100 Earth years) is calculated using expression

(3.1.1)
$$\frac{\varphi - \varphi_{kl}}{\varphi_{kl}} * 360 * 3600 * \frac{100}{0.2408},$$

the divider 0.2408 is used, because the Mercury year lasts 0.2408 Earth years. In equation 3.1.1 φ_{kl} is the azimuth angle calculated using Newton theory and φ is the azimuth angle calculated using Einstein theory. In Lawden book /2/ an expression is given for the relative perihelion shift (equation 53.22). It may be written using this book symbols for quantities as

(3.1.2)
$$\frac{\varphi - \varphi_{kl}}{\varphi} = \frac{3\alpha(r_1 + r_2)}{4 r_1 r_2} \quad \text{or} \quad \frac{\varphi - \varphi_{kl}}{\varphi_{kl}} = \frac{3\alpha(r_1 + r_2)}{4 r_1 r_2 - 3\alpha(r_1 + r_2)}$$

In calculation cases 2 – 8 the mass of the Sun has been varied and then the planet year length is not 0.2408 Earth years. To keep things simple and comparison easy the perihelion shift has been calculated using equation 3.1.1 also in these cases, but the value for the shift doesn't represent a value in 100 Earth years. The orbit calculation result is given in appendix 2. At the beginning of the output there have been written some input quantities and calculated constants. Print of the actual calculation result for the orbit begins after the heading 'calculation result'. In first column the angle got by numerical integration of expression 2.1.6 is given, so this is the angle given by the relativity theory. In second column the orbit radius of Mercury is given at the angle given in first column. The radius that is calculated using the ellipse equation

(3.1.3)
$$r = \frac{b^2}{a + c * cos\varphi}$$

is given in the third column (φ is the angle in the first column and c is the distance between the Sun and the central point of the orbit ellipse). The difference of the value in the second column and of the value in the third column is written in the fourth column. So the value in the fourth column gives the difference of the planet distance between the relativistic and classic calculations at the same azimuth angle. In the second row of the first column the result of numerical integration of equation 2.1.7 is given. This is the azimuth angle given by classical mechanics. So in column 1 at rows 1 and 2 there is given the azimuth angles according to relativity theory and according to classic mechanics at the same distance from the Sun. Integral perihelion shift calculated using formula 3.1.1 is shown in the second column and the differential (local) perihelion shift is shown in the third column

Relativistic and classic Mercury orbits are shown in figure 3.1.1. Schwarzschild radius is also drawn in the picture, but it is too small to be seen.

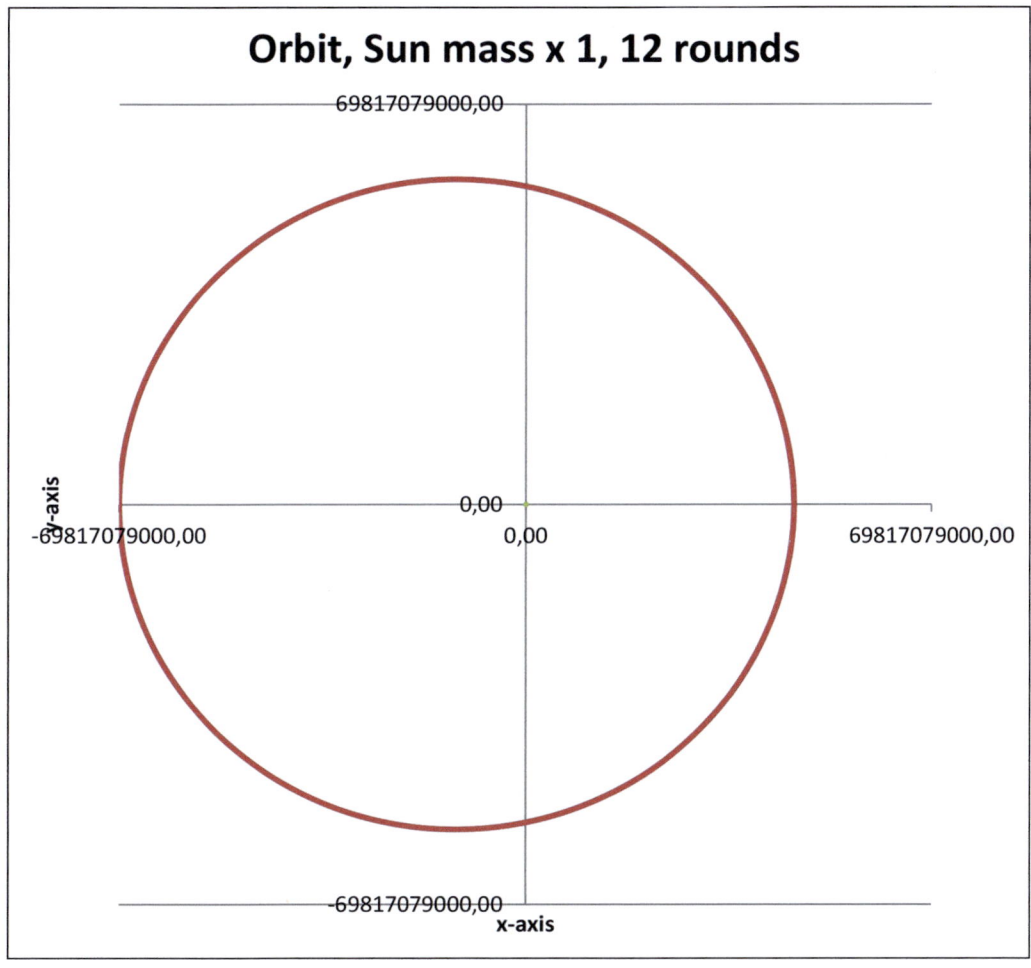

Figure 3.1.1: Relativistic and classic Mercury orbit.

Classic and relativistic orbits are completely overlapping with the resolution of figure 3.1.1. Perihelion shift as a function of angle is shown in figure 3.1.2. The differential shift seems to be a periodic function of angle and has the same value at the same angle in every round trip. Integral shift is also periodic, but it converges towards the value 43.0037 in case of Mercury orbit as the number of round trips increases. The integral shift has value 43.0037 at aphelion and perihelion in every round trip; this is the very same value toward which it converges as the number of round trips increases

The relative shift is calculated by dividing value 43.0037 with 360*3600*100/0.2408 = 5.382E8 and the result is 7.9E-8. Formula 3.1.2 gives the same result 7.9E-8.

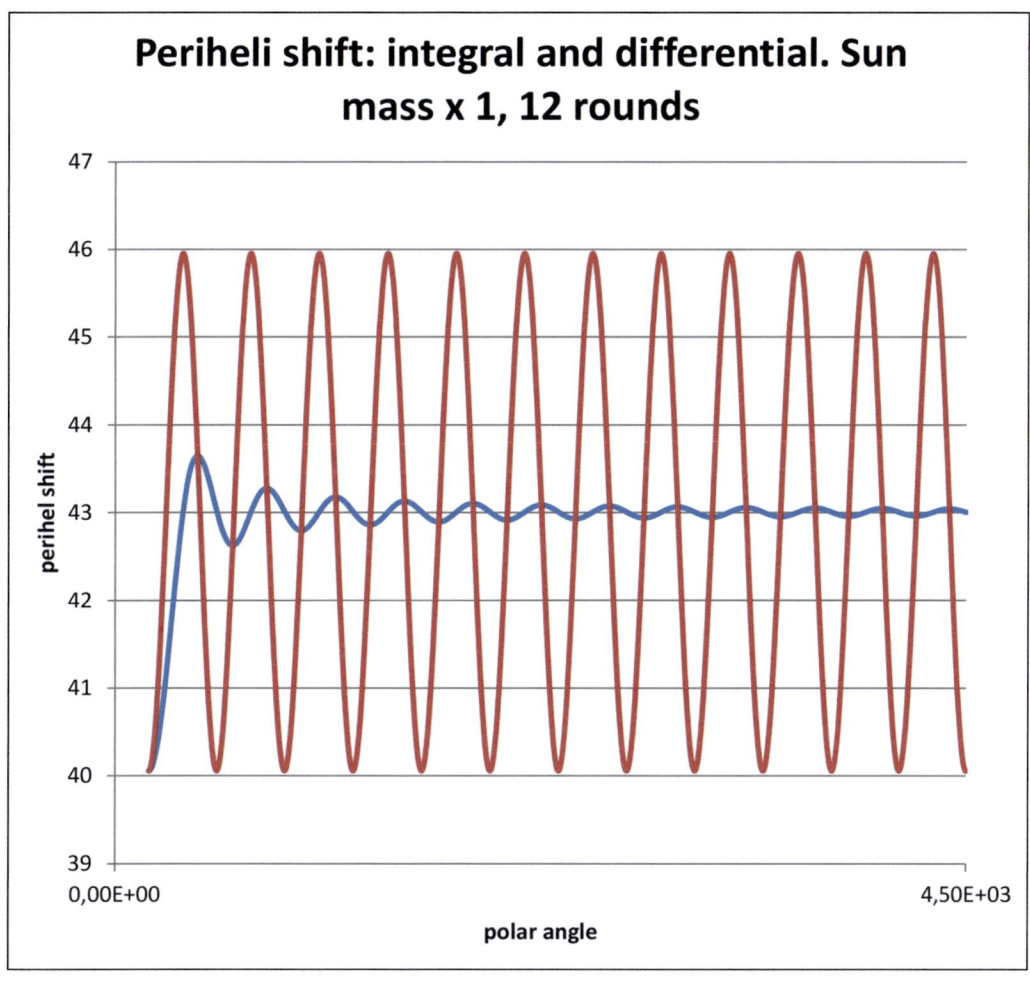

Figure 3.1.2: Perihelion shift (arc seconds in 100 Earth years), the calculation is started from aphelion

The calculation is started from aphelion in figure 3.1.2. The same calculation is started from perihelion in figure 3.1.3.

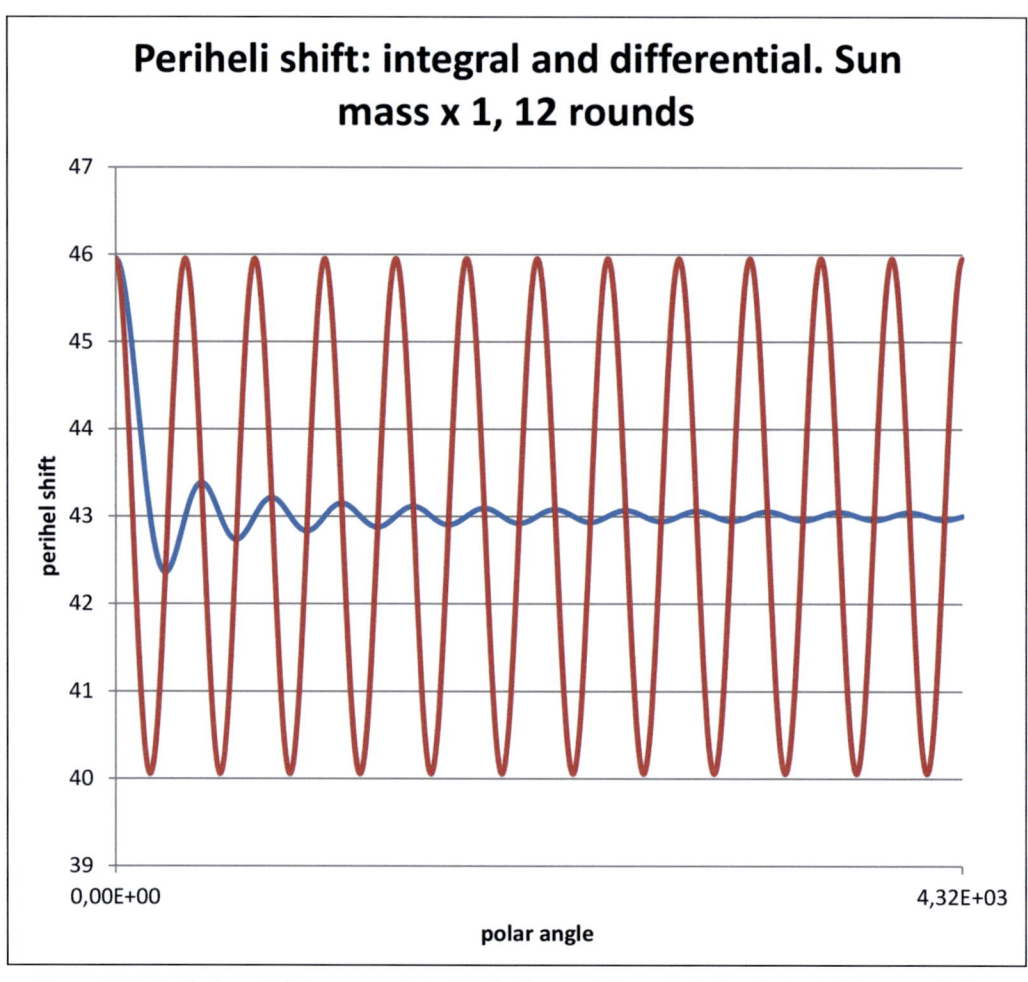

Figure 3.1.3: Perihelion shift (arc seconds in 100 Earth years), the calculation is started from perihelion

Both differential and integral perihelion shift behave logically in the same way as calculation is started either from perihelion or from aphelion. When Mercury is in aphelion or in perihelion the integral perihelion shift has value 43.0037 that is the same value towards which it converges as the number of round trips is increased.

In the following calculation cases 2 – 8 the effect of varying the Sun (or black hole) mass is examined. In all cases calculation method 1 is used. In all calculation cases parameters

$\bar{a}$ and $\bar{b}$ have been given the classical values that is the half value of the Mercury major axis and the half value of the Mercury minor axis

(3.1.4) $\bar{a} = a = 5.79091755E10$

(3.1.5) $\bar{b} = b = 5.66716379E10$

The Sun mass has been varied in the different calculation cases; it is multiplied by 0.4E6, 1.0E6, 2.522090E6, 2.522091E6, 3.0E6, 30E6 and 300E6. The corresponding Schwarzschild radius has been calculated and the perihelion and aphelion distances from the Sun have been determined by solving the roots of the 4-degree polynomial

(3.1.6) $f(r) = r\left(2\bar{a}r^2 - (\bar{b}^2 + r^2)(r - \alpha)\right) = 0.$

3.2 Calculation case 2: The Sun mass multiplied by 0.4E6

In this case the mass of the Sun is multiplied by 0.4E6. Six round trips around the Sun have been calculated. Calculation method 1 has been used (variable MOD has been given value = -1), so that values for parameters $\bar{a}$ and $\bar{b}$ have been calculated using expressions 3.1.4 and 3.1.5. The planet distance from the Sun at perihelion and aphelion are determined by solving equation 3.1.6. The solution gives values $r_1 = 4.1095E10$ and $r_2 = 7.46681E10$. Orbit equation $r = r(\varphi)$ is determined by numerical integration of expression 2.1.6 from value r_2 to value r_1.

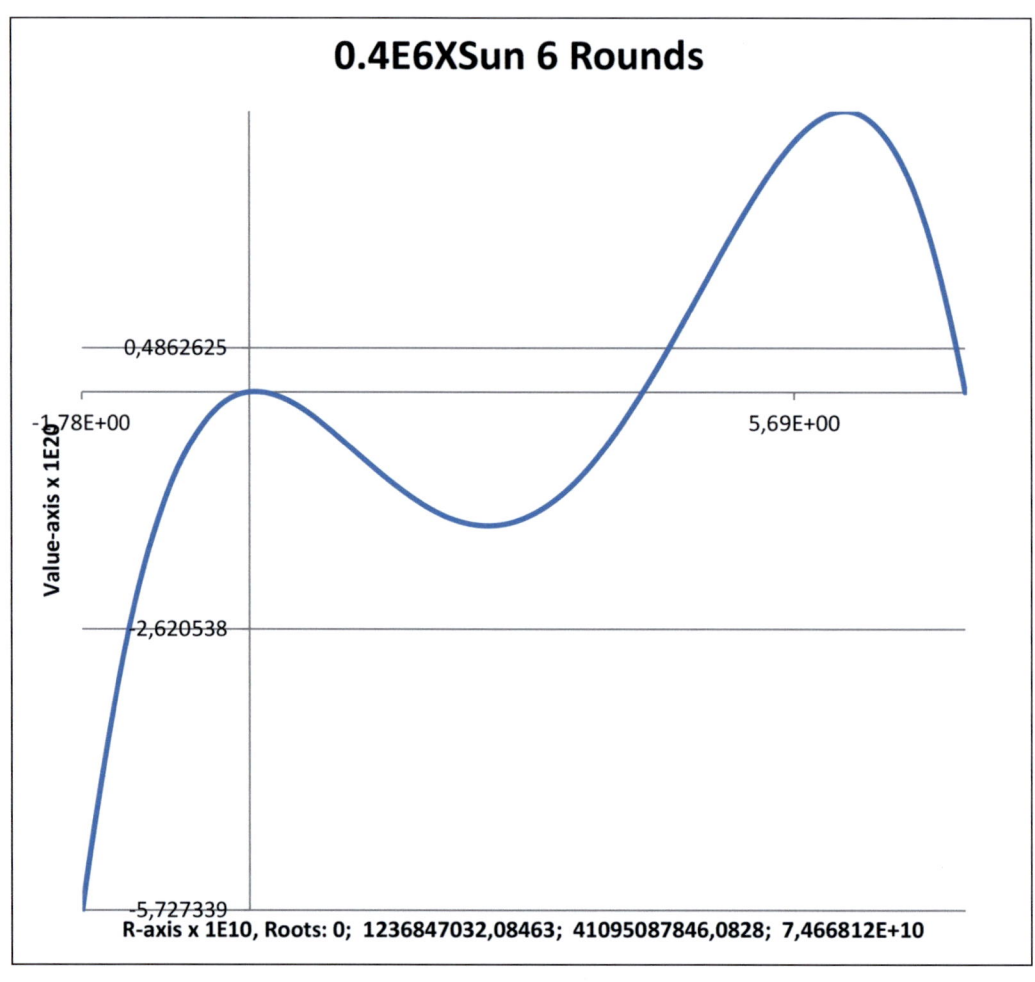

0.4E6XSun 6 Rounds

Value-axis x 1E20

0,4862625

-1,78E+00

5,69E+00

-2,620538

-5,727339

R-axis x 1E10, Roots: 0; 1236847032,08463; 41095087846,0828; 7,466812E+10

Figure 3.2.1: The graph of the 4-degree polynomial and values for its zero points, as the mass of the Sun has been multiplied by 0.4E6. The corresponding Schwarzschild radius is 1.18E9.

The Mercury orbit according to classical mechanics and according to relativity theory is drawn in figure 3.2.2. The orbit 'ellipse' according to relativity theory is circulating with a rather high speed. Circle at Schwarzschild radius is also drawn in the picture.

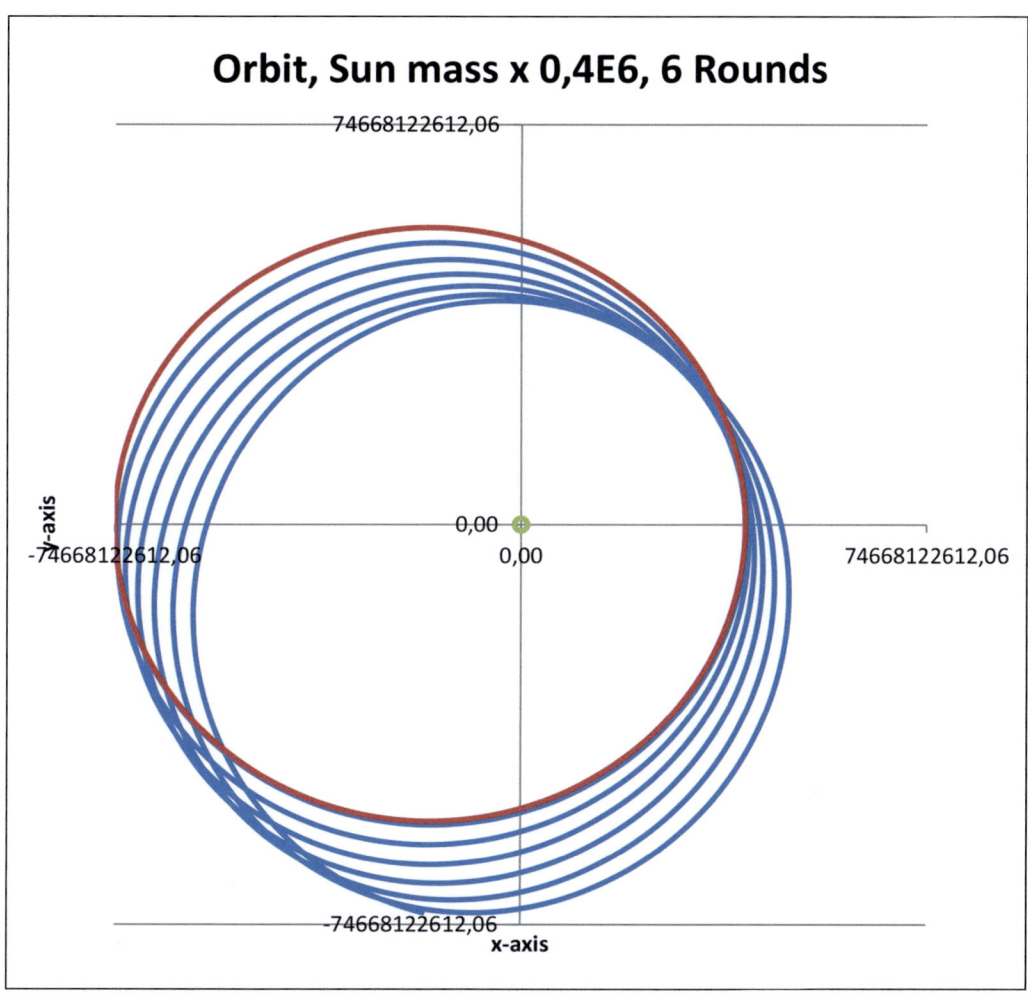

Figure 3.2.2: Planet orbit according to classical mechanics and according to relativity theory as the mass of the Sun is multiplied by 0.4E6. A little circle at Schwarzschild radius can be seen around the origin.

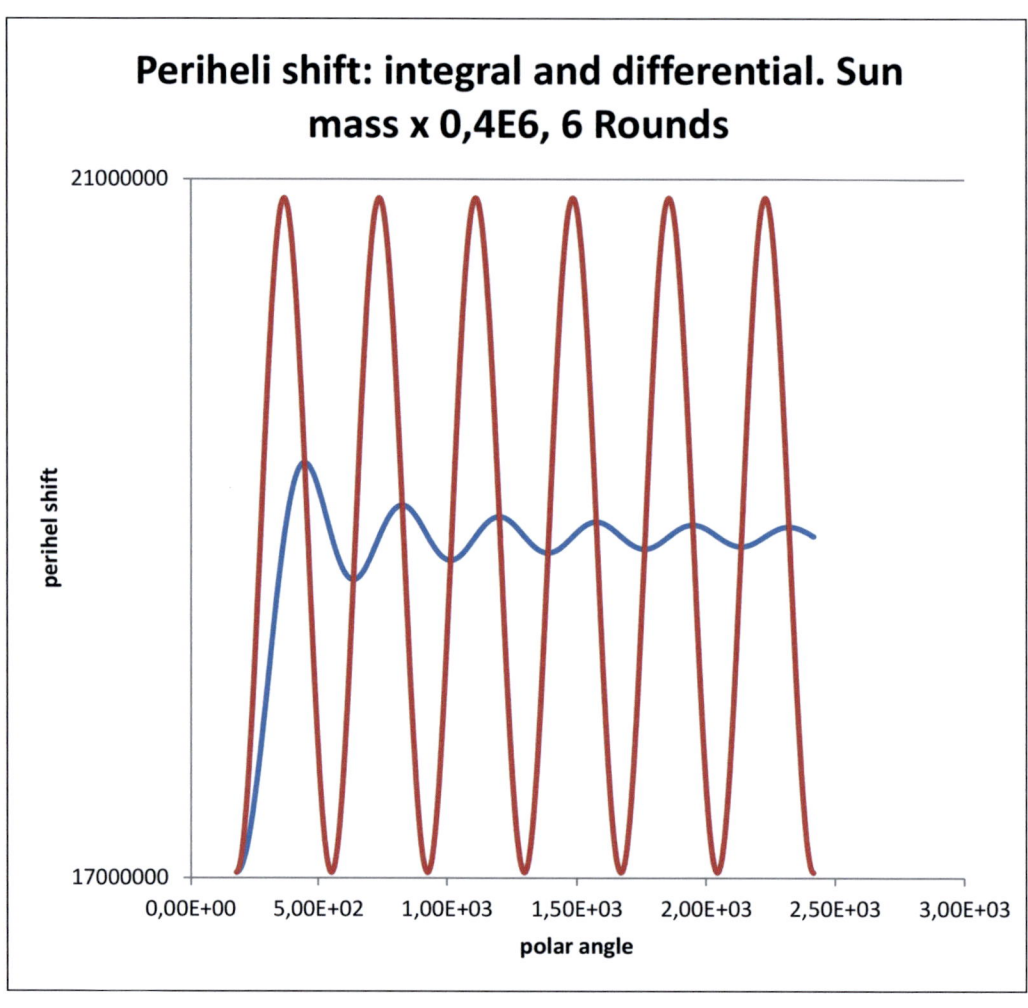

Figure 3.2.3: Perihelion shift calculated using equation 3.1.1, the value is 1.90E7.

The program calculates value 1.90E7 for the perihelion shift. The relative shift is calculated by dividing 1.90E7 with the number 360*3600*100/0.2408 = 5.382E8 and the result is 0.0353. Using equation 3.1.2 one gets a value 0.0346 for the perihelion shift. This value is about 2 % smaller than the value calculated by the program.

3.3 Calculation case 3: The Sun mass multiplied by 1.0E6

In this calculation the mass of the Sun is multiplied by the number 1.0E6. Three round trips around the Sun have been calculated. Calculation method 1 has been used (the variable MOD has been given the value -1), so that the constants $\bar{a}$ and $\bar{b}$ are calculated using expressions 3.1.4 and 3.1.5.

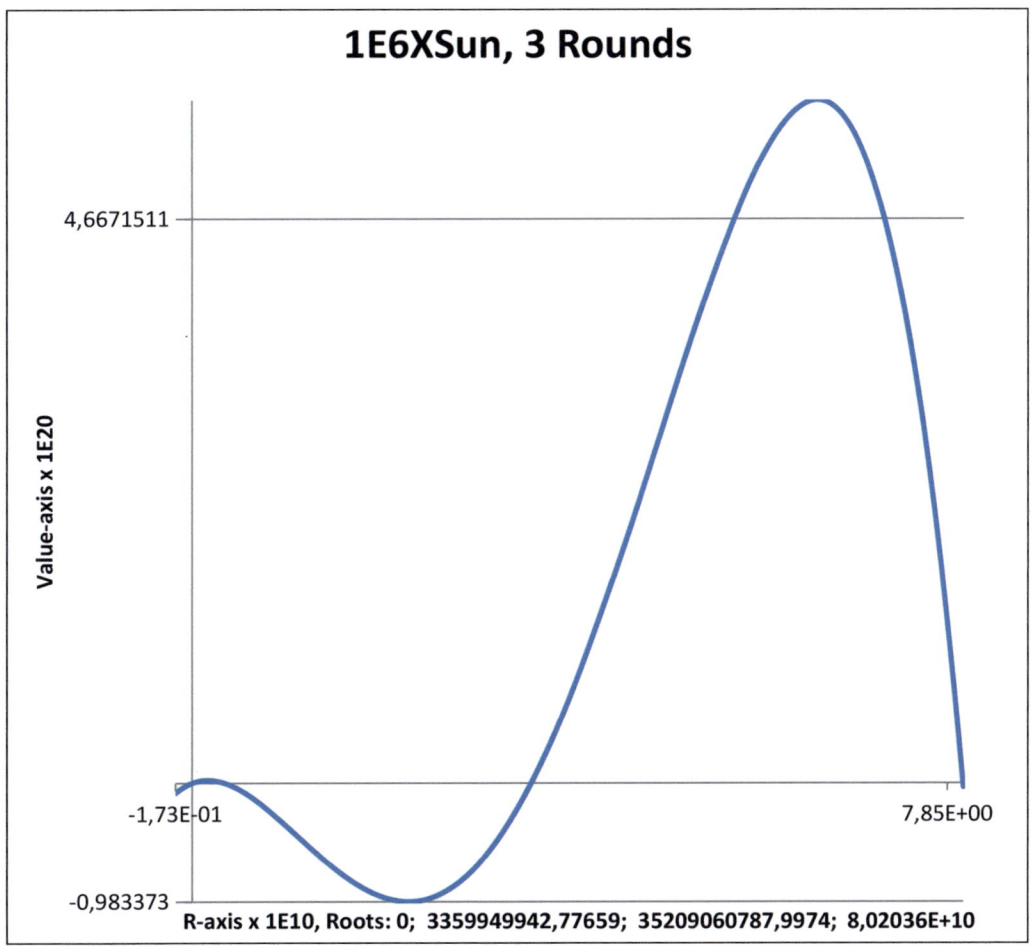

Figure 3.3.1: The graph of the 4-degree polynomial and the roots of the polynomial, as the mass of the Sun has been multiplied by 1.0E6. Now Schwarzschild radius gets the value 2.95E9.

31

The roots of the 4-degree polynomial are: $r_4 = 0$, $r_3 = 3.35995E9$, $r_1 = 3.52091E10$ and $r_2 = 8.02036E10$. The largest roots give the planet distance at perihelion and aphelion. Orbit equation $r = r(\varphi)$ is solved by numeric integration of expression 2.1.6 from value r_2 to value r_1. The planet orbit graph both according to relativity theory and according to classic mechanics is drawn in figure 3.3.2.

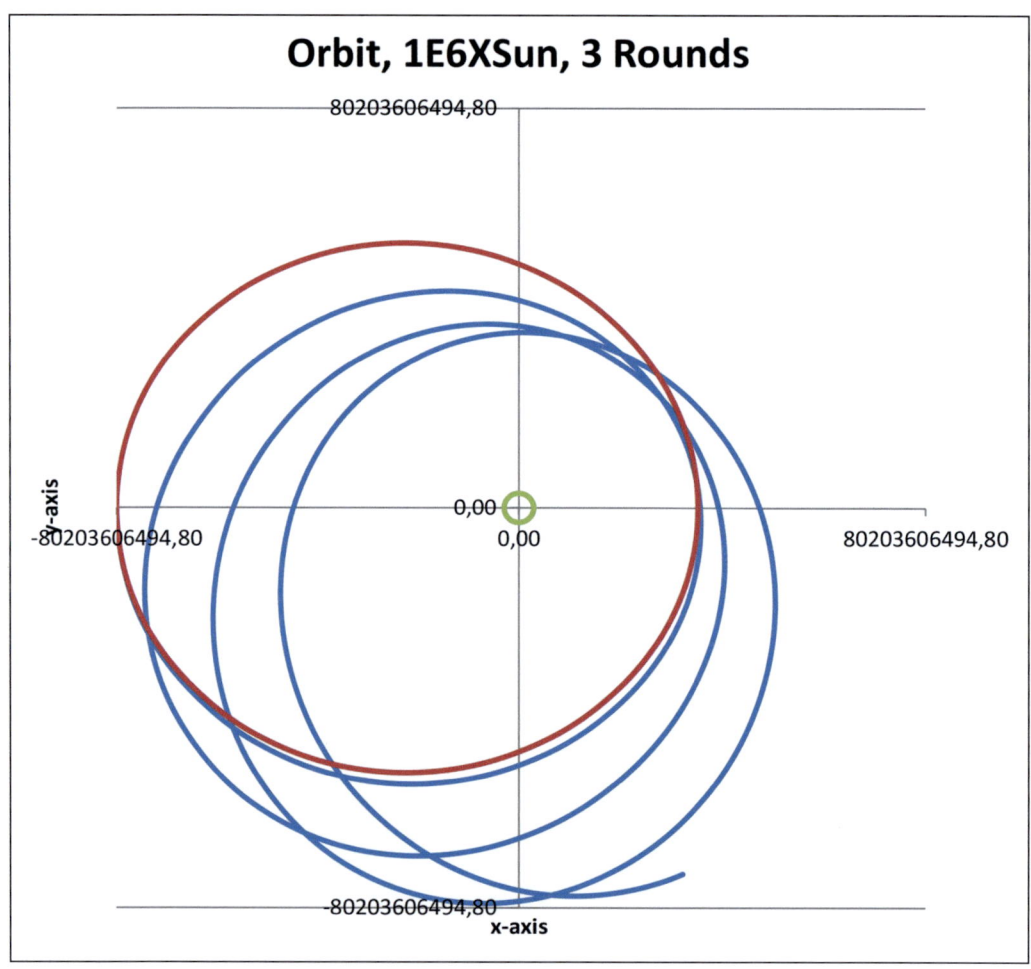

Figure 3.3.2: Planet orbit according to relativity theory and according to classic mechanics. A little circle at the Schwarzschild radius is drawn around the origin.

Classic orbit and relativistic orbit differ significantly from each other. If you have a good will, you may interpret the relativistic orbit as an ellipse rotating around the Sun. Perihelion shift is presented in figure 3.3.4 as a function of the polar angle coordinate and it is large. Local (differential) shift is a function of the polar angle. Also integral shift is a function of the polar angle, but it converges towards value 5.66E7, as the number of round trips around the Sun increases

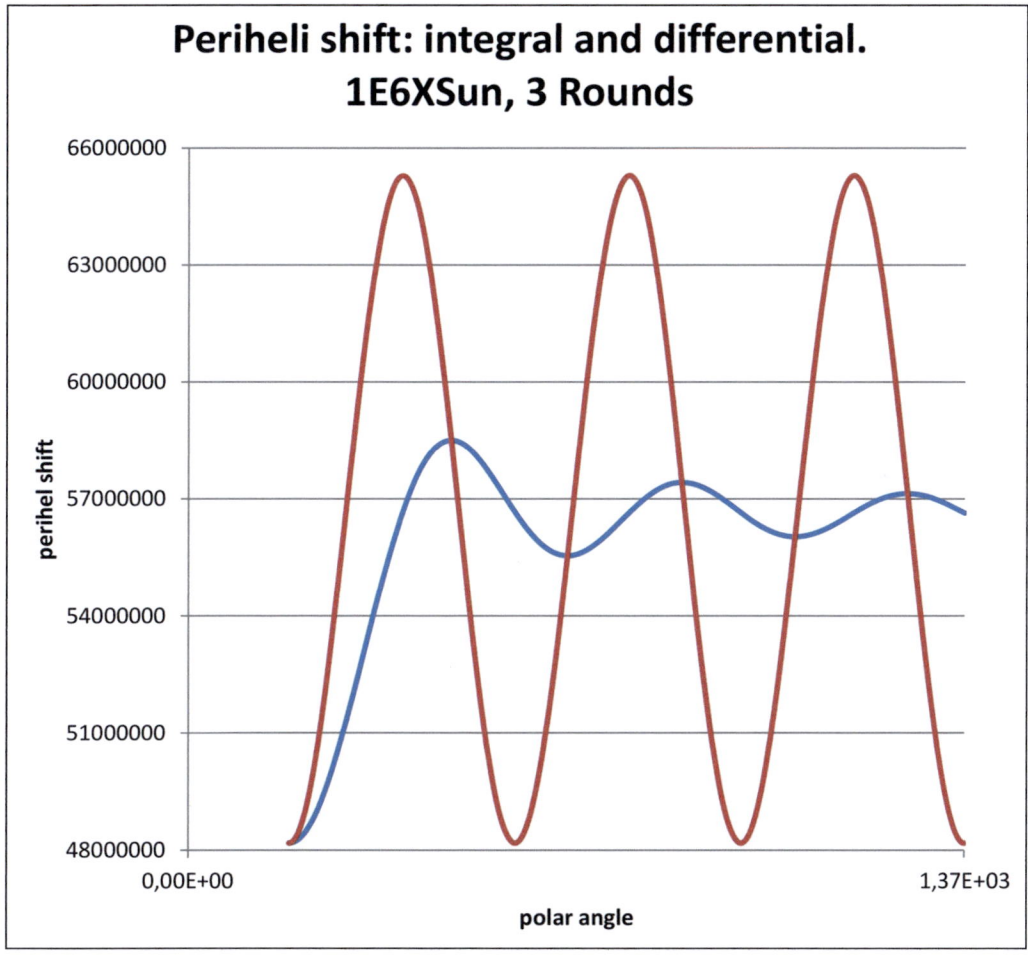

Figure 3.3.4: Integral perihelion shift calculated using equation 3.1.1

converges towards value 5.66E7 arc seconds.

The perihelion shift calculated by the program has the value 5.66E7. Relative shift is calculated by dividing value 5.66E7 by number 360*3600*100/0.2408 = 5.382E8 and the result is 0.1052. Expression 3.1.2 gives the result 0.0996, which is about 5 % smaller than the value calculated by the program.

3.4   Calculation case 4:
The Sun mass multiplied by 2.522090E6

In this calculation the mass of the Sun has been multiplied by number 2.522090E6. One round trip around the Sun has been calculated. With one round trip we mean the orbit from aphelion to aphelion, which actually means in the relativistic calculation about 4.9 round trips. The multiplier 2.522090E6 may appear arbitrary, but it has been chosen so that, if you increase the last decimal by 1, the planet falls into the black hole after about 4.9 round trips (look calculation case 5). We have used calculation method 1 (variable MOD in the program has given value -1), so that values for parameters $\bar{a}$ and $\bar{b}$ are given by expressions 3.1.4 and 3.1.5. Perihelion and aphelion distances from the Sun have been calculated by solving equation 3.1.6 and the values are $r_1 = 1.624E10$ and $r_2 = 9.08011E10$. The graph of the 4-degree polynomial is presented in figure 3.4.1. By looking the graph you may deduce that when the planet approaches the perihelion, the polar angle φ changes much even with small changes in r. In the orbit graph this may be seen so that the planet circles many times around the Sun at about the distance of the perihelion before it returns to aphelion. Orbit equation $r = r(\varphi)$ is calculated by numerical integration of expression 2.1.6 from value r_2 to value r_1. The orbit graph begins from aphelion and circulates about 5 round trips at about the perihelion distance, until it returns to aphelion and from there again to perihelion for about 5 round trips and so on.

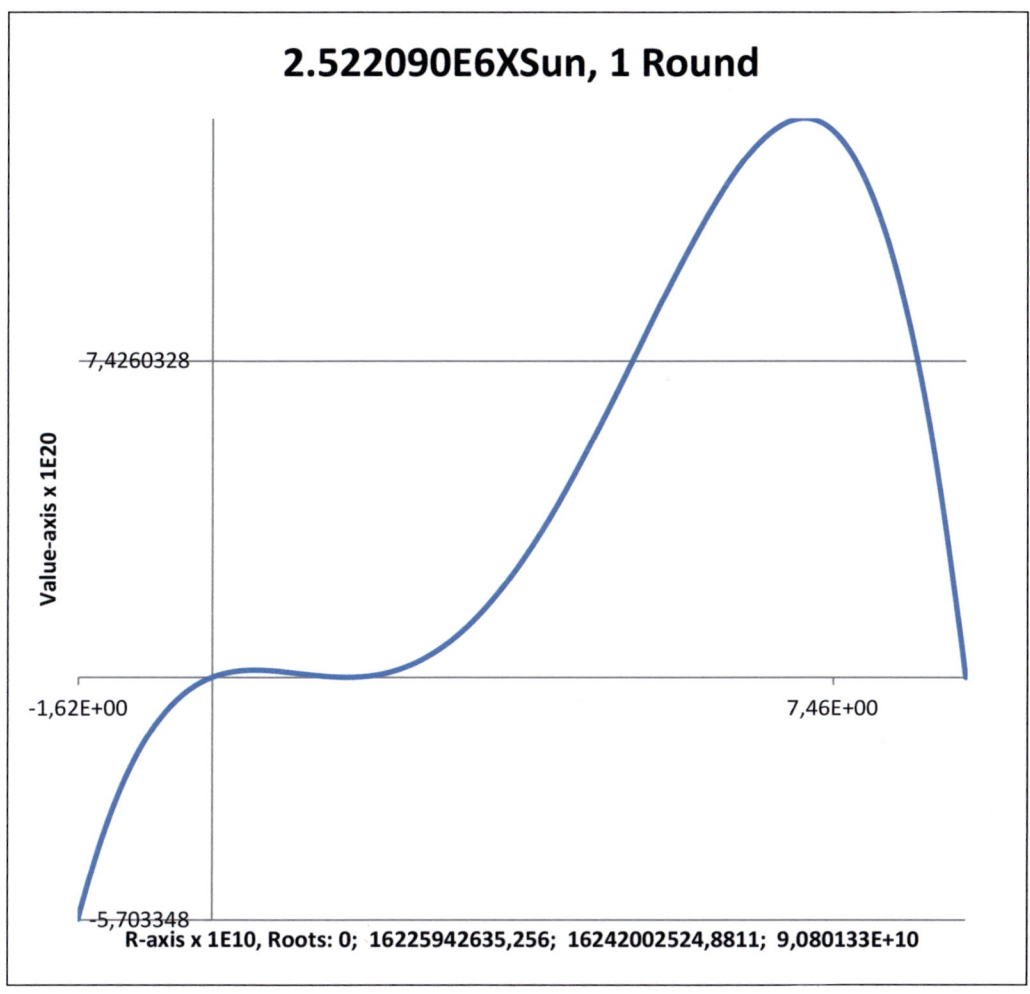

Figure 3.4.1: Graph and roots of the 4-degree polynomial, as the mass of the Sun has been multiplied by number 2,522090E6. Now Schwarzschild radius has value 7,4509253E9.

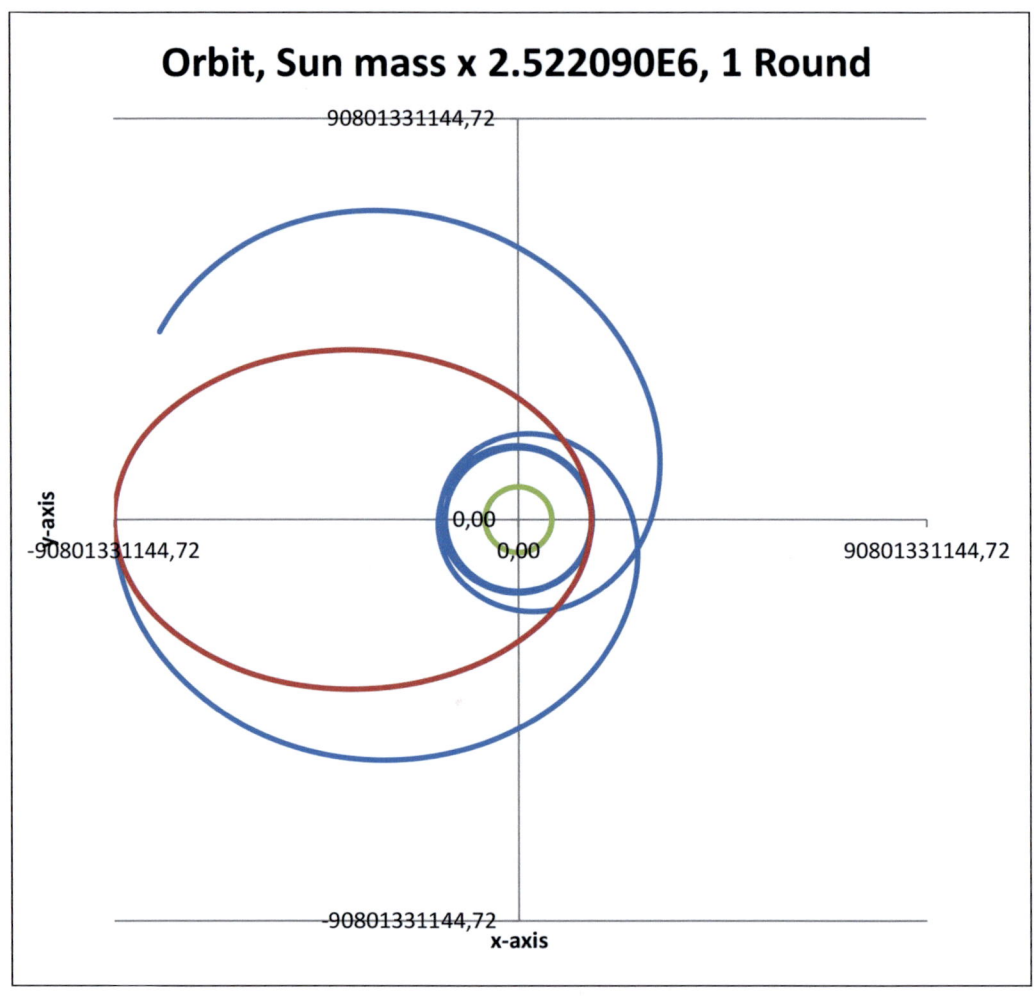

Figure 3.4.2: Classic and relativistic planet orbit. Planet circulates about 3 times around the Sun before it returns to aphelion.

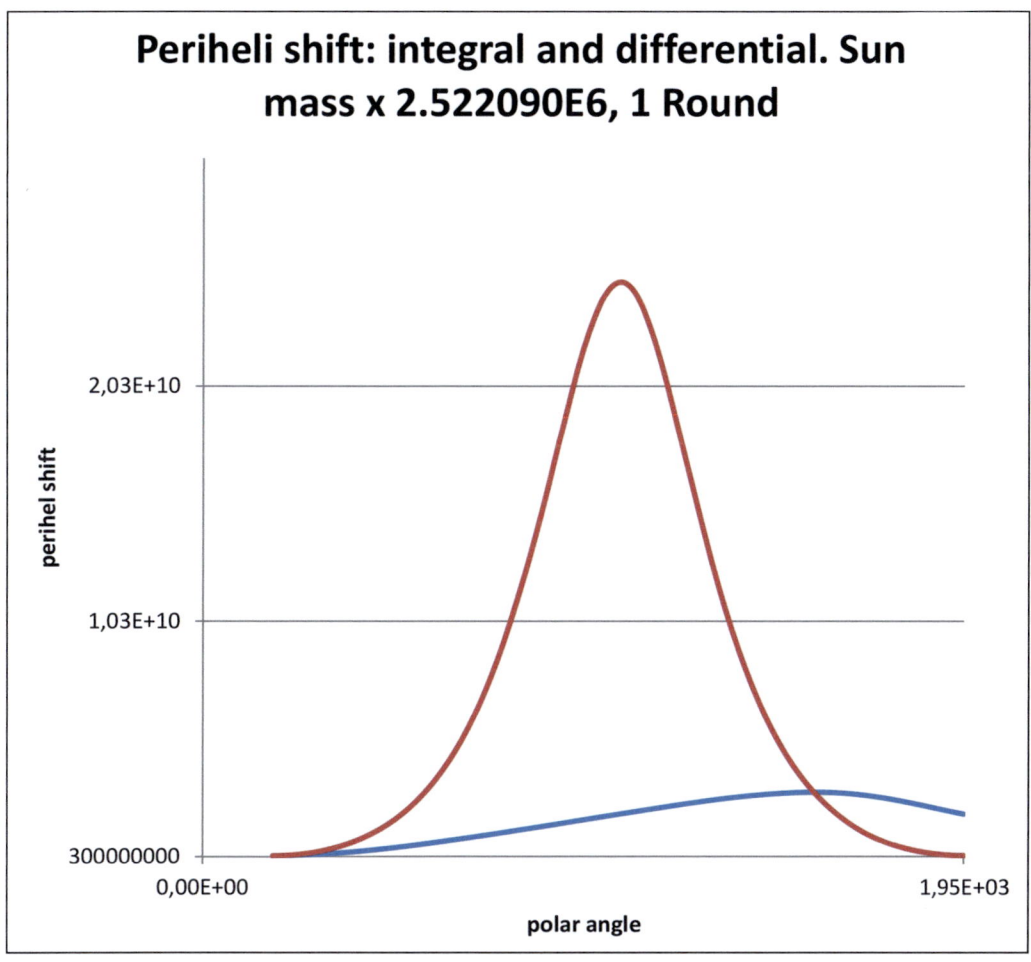

Figure 3.4.3: Integral perihelion shift calculated using equation 3.1.1 converges to value 2.11E9 arc seconds.

Program calculates value 2.11E9 for the perihelion shift. The relative shift is calculated by dividing 2.11E9 by number 360*3600*100/0.2408 = 5.382E8 and the result is 3.920. You get value 0.6824 for the relative shift by using equation 3.1.2. So the value calculated by the program is equal to the value given by equation 3.1.2 multiplied by number 5.7.

I solved the orbit equation also, as the Sun mass was multiplied by number

2.52209046831968488E6. In this case the planet circulates around the Sun about 11 round trips (4120 °) before it returns to aphelion. The largest number of round trips occurs at about the perihelion distance from the Sun. Apparently such an orbit that circulates around the Sun at about the perihelion distance for a very long time is possible. The planet orbit graph is drawn in figure 3.4.4.

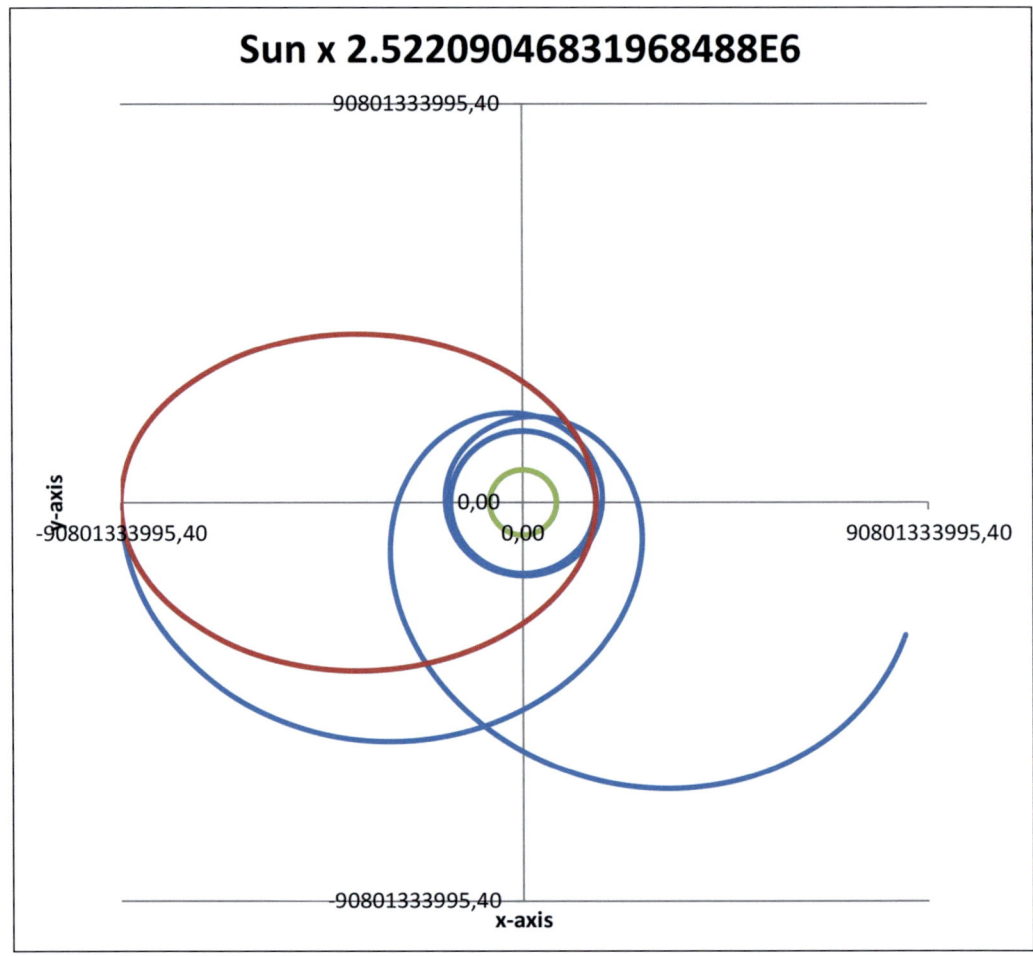

Figure 3.4.4 Classic and relativistic planet orbit, as the Sun mass has been multiplied by number 2.52209046831968488E6. Relativistic orbit circulates about 10 times at about perihelion distance before return to aphelion.

3.5 Calculation case 5:
The Sun mass multiplied by 2.522091E6

In this case the mass of the Sun has been multiplied by number 2.522091E6. 0.5 round trips have been calculated around the Sun. A half round trip is to be understood as the orbit from aphelion to perihelion, which in the relativistic calculation means about 4.7 round trips. Calculation method 1 (variable MOD = -1) has been used, so that parameters $\bar{a}$ and $\bar{b}$ are given in expressions 3.1.4 and 3.1.5. Perihelion and aphelion distances are the largest roots of equation 3.1.6 with values $r_1 = 0$ and $r_2 = 9.08011E10$. The graph of the 4-degree polynomial is drawn in figure 3.5.1. Now the polynomial has only 2 real roots and 2 complex roots. For the sake of simplicity the values of the complex roots are not given in figure 3.5.1, but three roots with value 0 are given. The smaller real root of the polynomial has value 0, which means that the planet falls into the black hole. The orbit equation $r = r(\varphi)$ is calculated by numerically integrating expression 2.1.6 from value r_2 to value r_1. The planet circulates a few times around the black hole, before it falls into the black hole. The largest number of round trips occurs with a radius on both sides the minimum of the polynomial; at that radius the change in polar angle is largest with a fixed change Δr of the distance r (look expression 2.1.6).

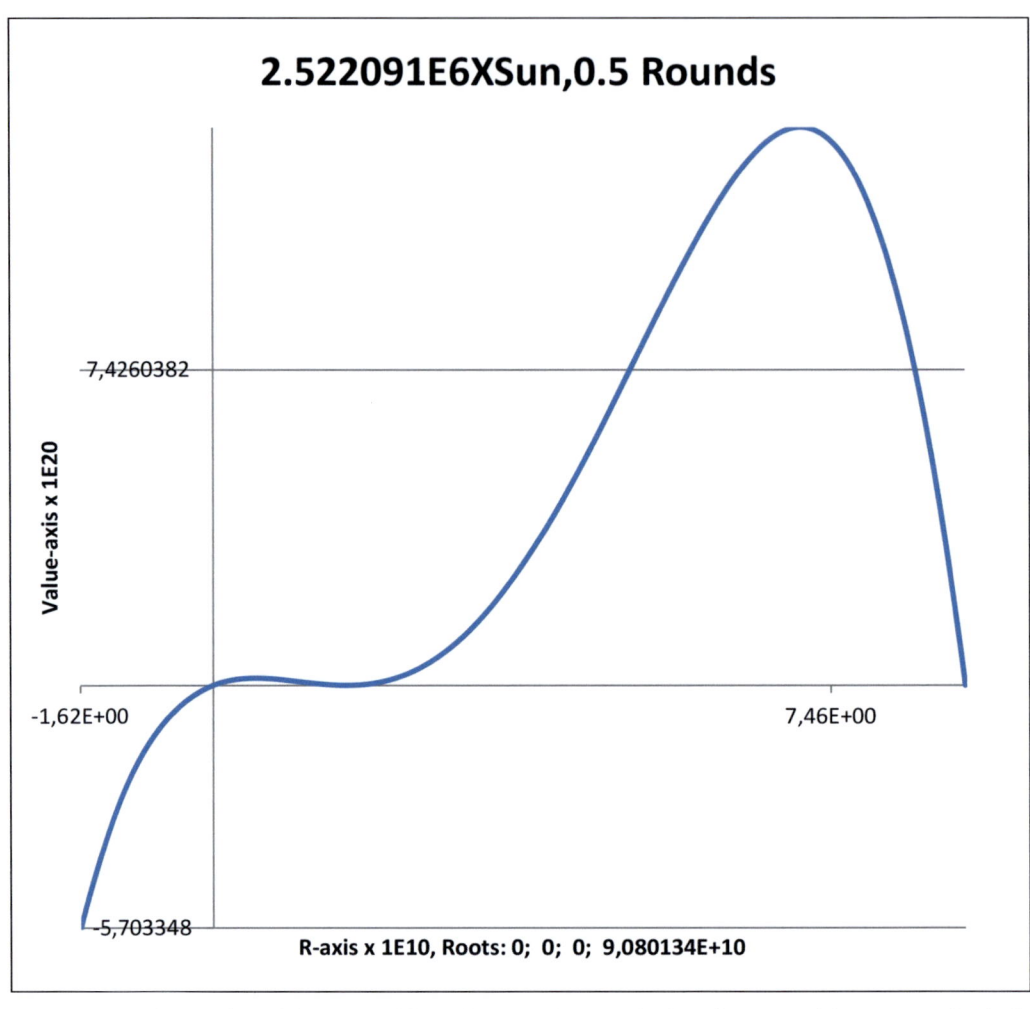

Figure 3.5.1: The graph and the roots of the 4-degree polynomial, when the mass of the Sun is multiplied by number 2.522091E6. In this case Schwarzschild radius has value 7,4509283E9.

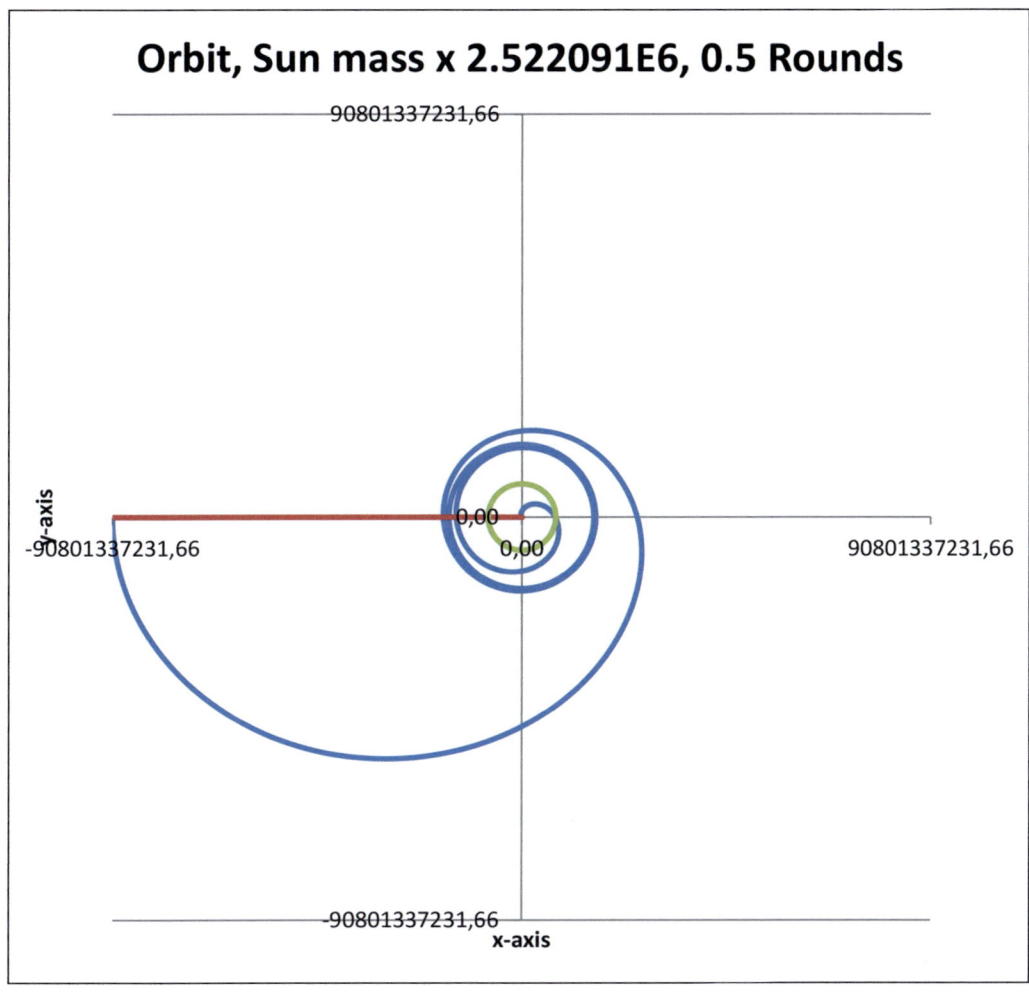

Figure 3.5.2: Classical and relativistic planet orbit. Planet circulates around the black hole a few times and then falls into the black hole. Classic orbit is a straight fall into the black hole.

3.6 Calculation case 6: The Sun mass multiplied by 3.0E6

In this calculation the Sun mass has been multiplied by the number 3.0E6. One round trip around the Sun has been calculated. A round trip is to be understood as the planet orbit from aphelion to aphelion, which in the relativistic calculation corresponds to about 2.6 round trips. Calculation method 1 is used (variable MOD has been given value -1 in the program), so that parameters $\bar{a}$ and $\bar{b}$ are given values in expressions 3.1.4 and 3.1.5. Perihelion and aphelion distances have been calculated by solving equation 3.1.6 and the result is $r_1 = 0$ and $r_2 = 9.3624E10$. Now the 4-degree polynomial has only 2 real roots and 2 complex roots. However for the simplicity three zero valued roots have been printed in figure 3.6.1 and values for the complex roots have not been given. The smaller real root of the polynomial has value 0, which means that the planet falls into the black hole singularity. This happens after about 1.3 round trips that is with less twisting than in the previous calculation case. The orbit continues out from the singularity back to aphelion. This is to be understood only as the graph of the orbit, the planet itself would move also in this part of the orbit towards the black hole singularity, as it has been commonly agreed that nothing comes out of a black hole. The orbit equation $r = r(\varphi)$ has been calculated by integrating expression 2.1.6 numerically from value $r = r_2$ to value $r = r_1$.

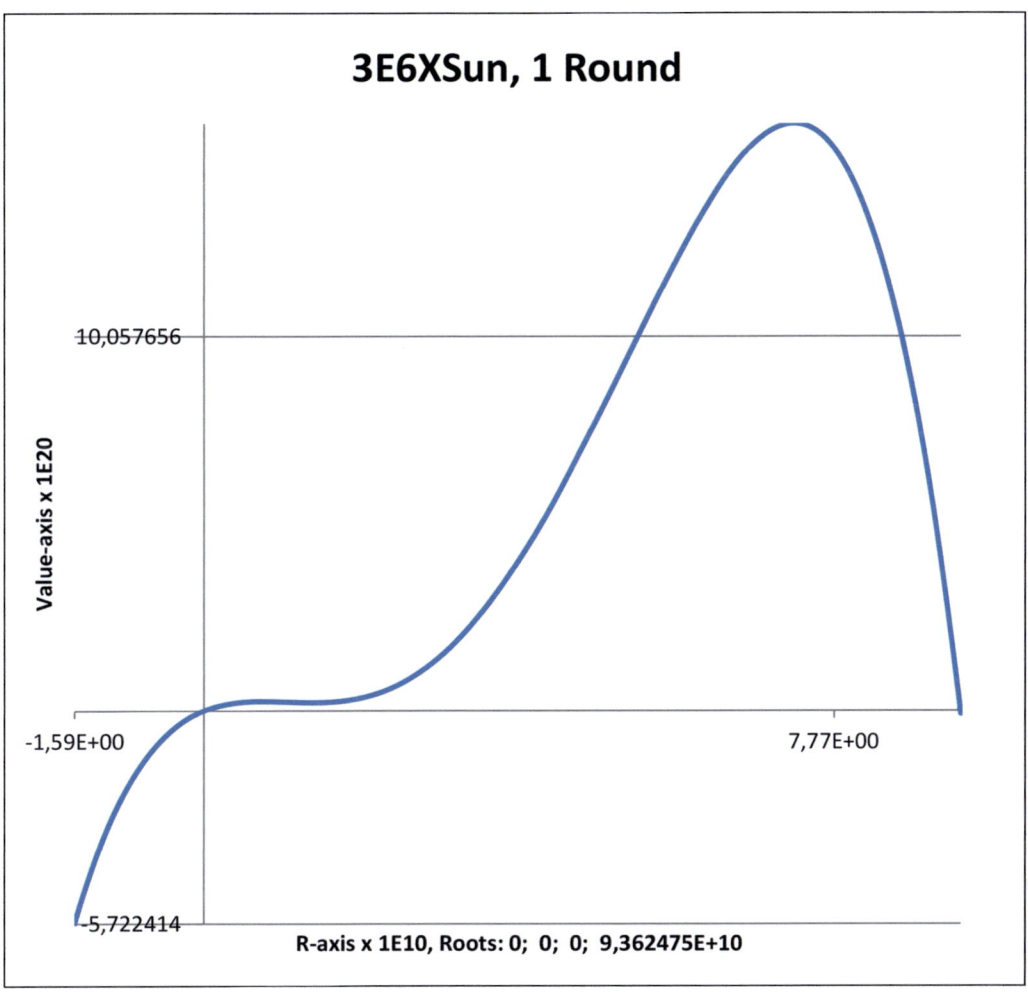

Figure 3.6.1: Graph and roots of the 4-degree polynomial, as the mass of the Sun has been multiplied by number 3.0E6. Now Schwarzschild radius has value 8,86E9.

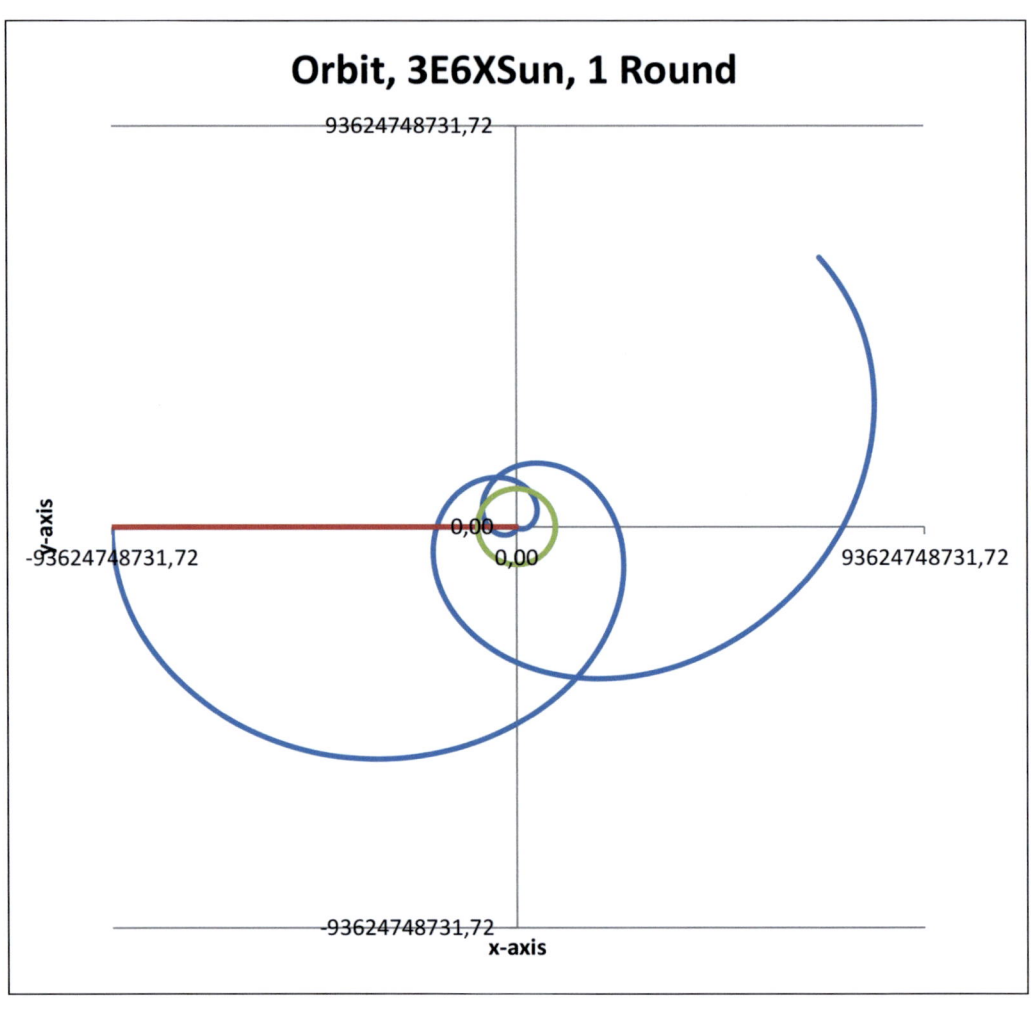

Figure 3.6.2: Classic and relativistic planet orbit. Classic orbit is a straight fall into the black hole, relativistic orbit twists a bit before the fall into the black hole centre (the continuation of the orbit out of the black hole describes only the orbit, the planet would travel also along this part of the orbit towards the black hole).

3.7 Calculation case 7: The Sun mass multiplied by 30.0E6

In this calculation the mass of the Sun has been multiplied by the number 30.0E6. Two round trips around the Sun have been calculated. A round trip is to be understood as the orbit from aphelion to aphelion, which corresponds to 0.8 round trips in the relativistic calculation. Calculation method 1 has been used (constant MOD has been given the value = -1), values for parameters $\bar{a}$ and $\bar{b}$ have been given in expressions 3.1.4 and 3.1.5. Perihelion and aphelion distances from the Sun are calculated by solving equation 3.1.6 and the solution is $r_1 = 0$ and $r_2 = 1.9547E11$. Now the polynomial has only two real roots and two complex roots. However for the simplicity three zero valued roots have been printed in figure 3.7.1 and values for the complex roots have not been given. The smaller real root of the polynomial has value 0, which means that the planet falls into the black hole singularity. This happens after about 0.4 round trips that is with less twisting than in the former calculation case. The orbit continues out from the singularity back to aphelion. This is to be understood only as the graph of the orbit, the planet itself would move also in this part of the orbit towards the black hole singularity, as it has been commonly agreed that nothing comes out of a black hole. The orbit equation $r = r(\varphi)$ has been calculated by integrating expression 2.1.6 numerically from value $r = r_2$ to value $r = r_1$ and then from value $r = r_1$ to value $r = r_2$.

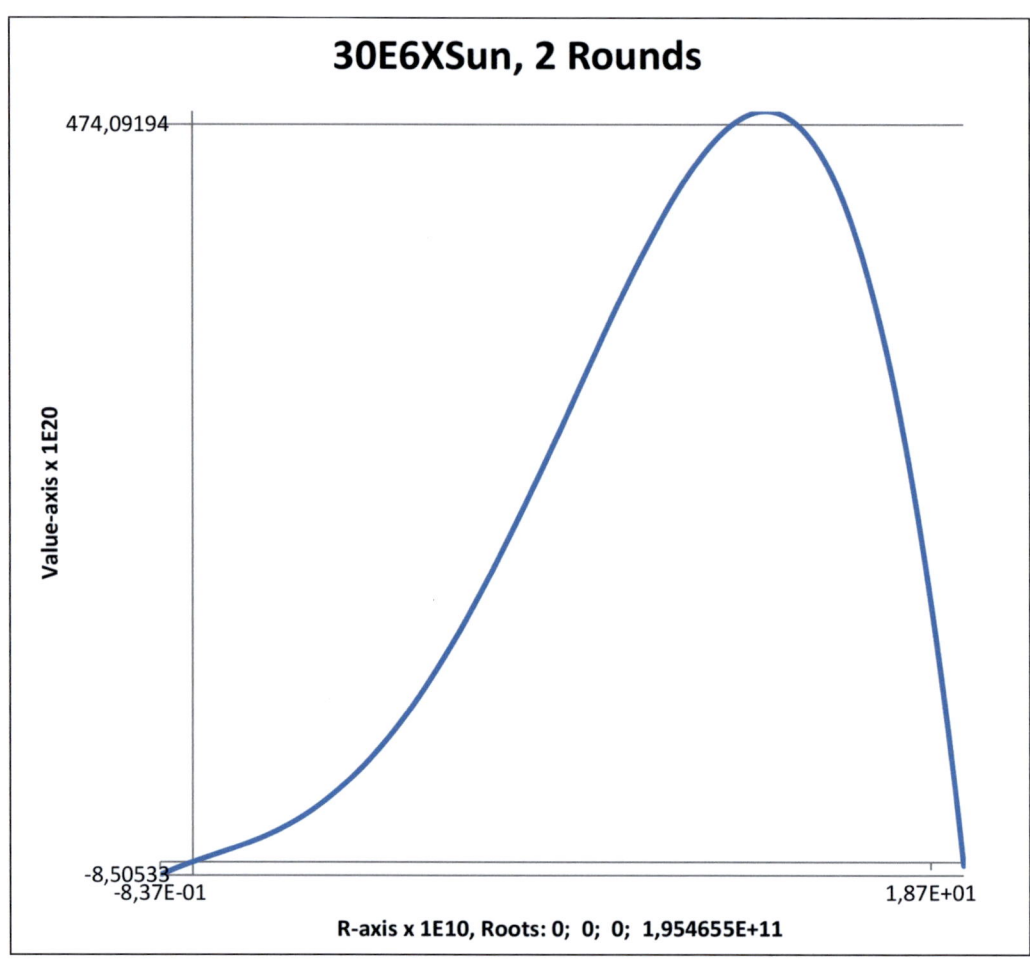

Figure 3.7.1: Graph and roots of the 4-degree polynomial, as the Sun mass has been multiplied by number 30.0E6. Now Schwarzschild radius has value 8,86E10.

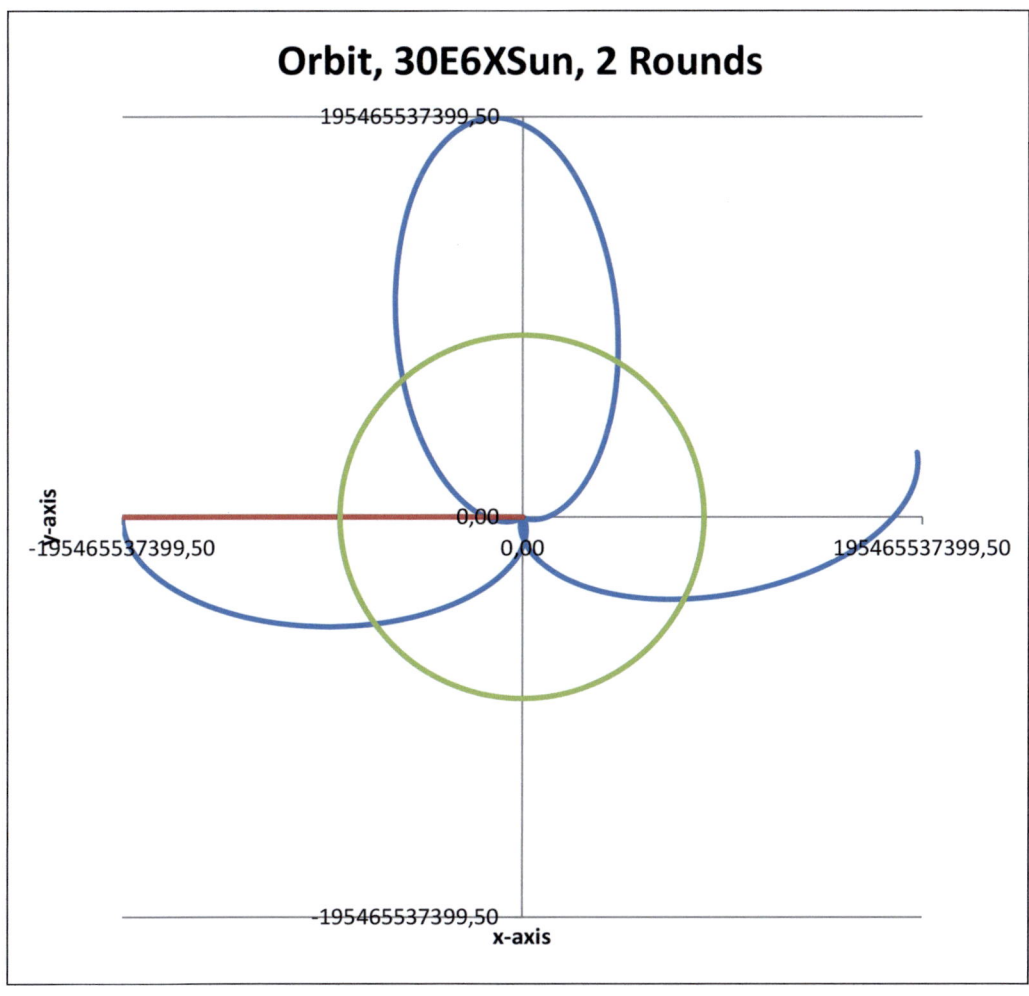

Figure 3.7.2: Classic and relativistic planet orbit. Classic orbit is a straight fall into the black hole, relativistic orbit twists a bit before falling into the centre of the black hole (the orbit continues out of the black hole but the planet would move also along this part of the orbit towards the black hole centre).

3.8 Calculation case 8: The Sun mass multiplied by 300.0E6

In this calculation the mass of the Sun has been multiplied by the number 300.0E6. Two round trips around the Sun have been calculated. A round trip is to be understood as the orbit from aphelion to aphelion, which corresponds to 0.3 rounds in the relativistic calculation. Calculation method 1 has been used (constant MOD has been given the value = -1), values for parameters $\bar{a}$ and $\bar{b}$ have been given in expressions 3.1.4 and 3.1.5. Perihelion and aphelion distances from the Sun are calculated by solving equation 3.1.6 and the solution is $r_1 = 0$ and $r_2 = 1.0017E12$. Now the polynomial has only two real roots and two complex roots. However for the simplicity three zero valued roots have been printed in figure 3.8.1 and values for the complex roots have not been given. The smaller real root of the polynomial has value 0, which means that the planet falls into the black hole singularity. This happens with less twisting than in the former calculation case. The orbit continues out from the singularity back to aphelion. This is to be understood only as the graph of the orbit, the planet itself would move also along this part of the orbit towards the black hole singularity, as it has been commonly agreed that nothing comes out of a black hole. The orbit equation $r = r(\varphi)$ has been calculated by integrating expression 2.1.6 numerically from value $r = r_2$ to value $r = r_1$ and then from value $r = r_1$ to value $r = r_2$.

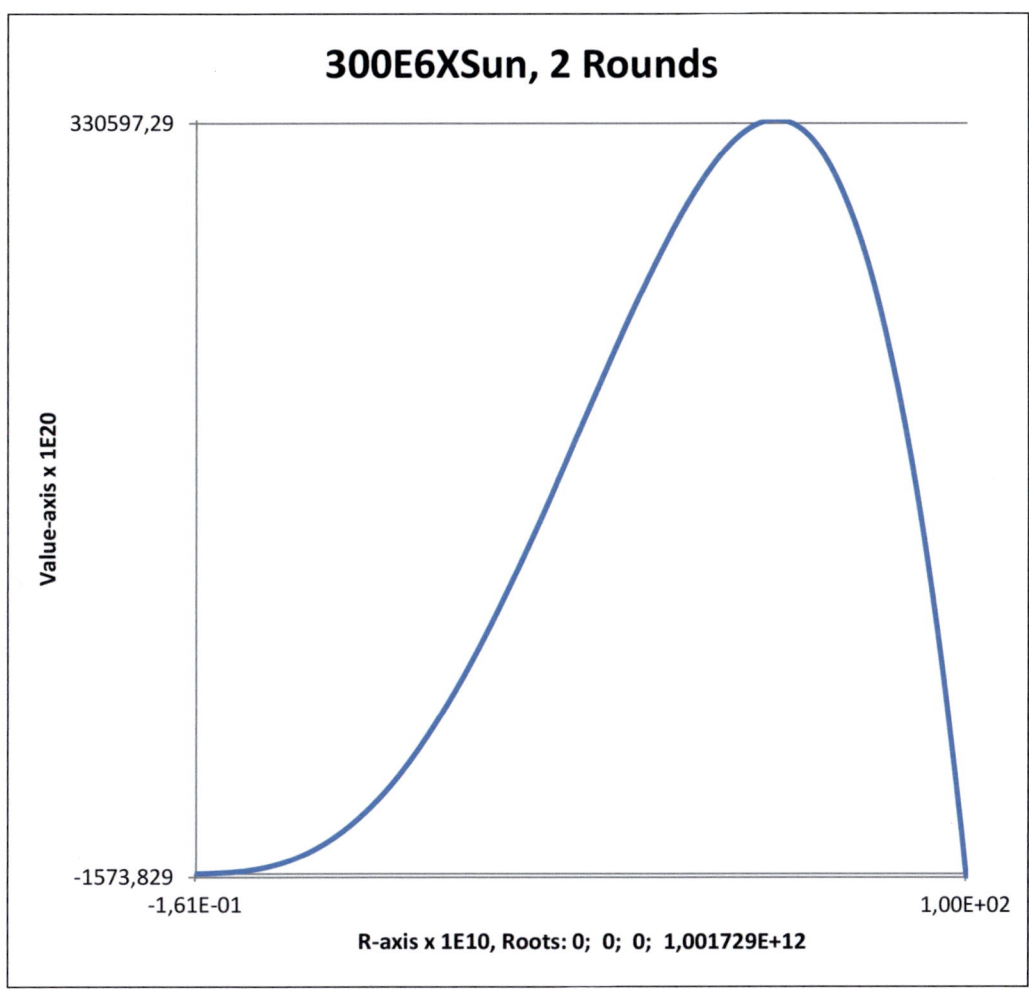

Figure 3.8.1: Graph and roots of the 4-degree polynomial, as the mass of the Sun has been multiplied by the number 300.0E6. Now Schwarzschild radius has value 8,86E11.

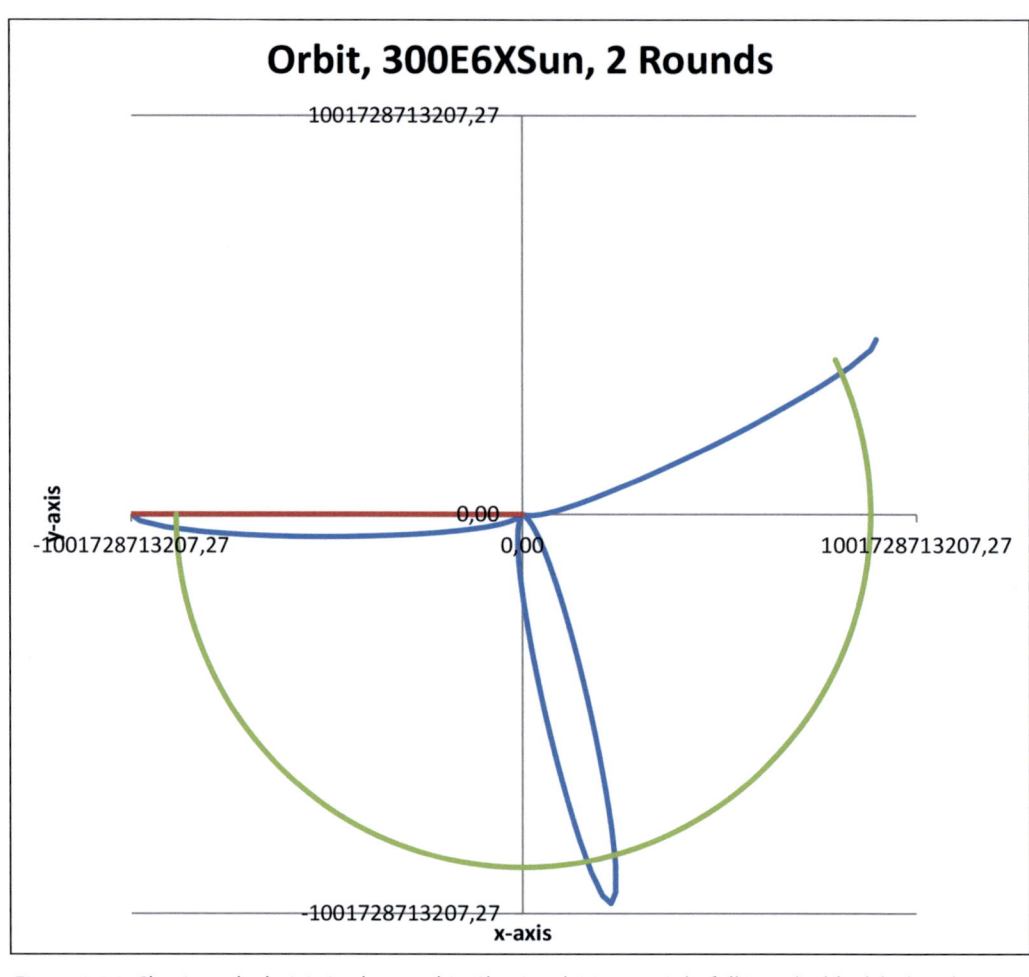

Figure 3.8.2: Classic and relativistic planet orbit. Classic orbit is a straight fall into the black hole, relativistic orbit twists a bit before falling into the centre of the black hole (the orbit continues out of the black hole but the planet would move also along this part of the orbit towards the black hole centre).

4   ORBIT EQUATION IN THE LITERATURE

4.1   Introduction

This book starts with the Euler equations presented in reference /1/. Normal way, which is presented in the literature, to construct the Euler equations is different from the way presented in reference /1/. Integrating the 'normal' equations one gets an orbit equation, which includes two integration constants. One may derive values for these constants by demanding that the relativistic orbit equation approaches the classic orbit equation in the classic limit (speed of light $c \rightarrow \infty$), but mathematically there could be a multiplying constant that in the classic limit approaches value 1. In reference /1/ and in this book the Euler equations are constructed in a different way compared to the normal way. Demanding that the planet distance from the Sun in perihelion and aphelion is same in classic and relativistic calculations, one may derive such values for the integration constants as functions of the perihelion and aphelion distances and Schwarzschild radius, which approach in the classic limit the corresponding classic parameter values (half of the orbit ellipse major and minor axes).

4.2   A typical orbit equation calculation

A typical way to construct the orbit equation is presented in Lawden book /2/. This way to construct the orbit equation can be found in many books dealing with the gravitation theory.

We shall proceed as in reference /2/. Integrating Euler equations (reference /2/ 53.4 and 53.5 page 147) one ends up with the equations (reference /2/ 53.7 and 53.8, page 148)

(4.2.1)
$$\frac{d\Phi}{d\tau} = \frac{h}{r^2},$$

(4.2.2)
$$\frac{dt}{d\tau} = \frac{kr}{r-\alpha}$$

where h and k are integration constants. In the above equations constant m has been given value α/2, where

$$\alpha = \frac{2M\bullet}{c^2}$$

is the Schwarzschild radius. This connection has been given in Lawden book. φ is the azimuth angle and r is the planet distance from the Sun.

Derivation has been done in respect to the local proper time τ

(4.2.3) $\quad d\tau^2 = -\frac{1}{c^2}\left(\frac{dr^2}{1-\frac{\alpha}{r}} + r^2(d\theta^2 + sin^2\theta d\varphi^2)\right) - \left(1 - \frac{\alpha}{r}\right)c^2dt^2$

According to the terminology used in reference /1/ and in this book $\quad \tau = \frac{s}{c}$ is the four dimensional distance s divided by the speed of light c.

A differential equation for the planet distance from the Sun is derived in Lawden book (reference /2/ equation 53.10, page 148)

(4.2.4) $\qquad r_\varphi^2 = \frac{(k^2-1)c^2}{h^2}r^4 + \frac{\alpha c^2}{h^2}r^3 - r^2 + \alpha r$

Make change of variable $\quad u = \frac{1}{r}.$
and you get (reference, /2/ equation 53.11, page 148)

(4.2.5) $\qquad u_\varphi^2 = \frac{(k^2-1)\, c^2}{h^2} + \frac{\alpha c^2}{h^2}u - u^2 + \alpha u^3$

Differentiate with respect to φ and you get (reference /2/, equation 53.12, page 148)

(4.2.6) $\qquad u_{\varphi\varphi} + u = \frac{\alpha c^2}{2h^2} + \frac{3}{2}\alpha u^2$

We shall now compare the equations of Lawden book to the equations of this book. According to equations 4.2.1 and 4.2.2

(4.2.7) $\qquad \frac{h}{r^2} = \frac{d\varphi}{d\tau} = \frac{d\varphi}{dt}\frac{dt}{d\tau} = \frac{d\varphi}{dt}\frac{kr}{r-\alpha}$

so that

(4.2.8) $\qquad \frac{d\varphi}{dt} = \frac{h}{k}\frac{r-\alpha}{r^3}$

Compare with equation 2.1.3 and you get

(4.2.9)
$$h = k\bar{b} \sqrt{\frac{GM}{\bar{a}}}$$

Insert expression 4.2.9 into equations 4.2.4, 4.2.5 and 4.2.6. The result is

(4.2.10)
$$r_\varphi^2 = \frac{2\bar{a}(k^2-1)}{k^2 \alpha \bar{b}^2} r^4 + \frac{2\bar{a}}{k^2 \bar{b}^2} r^3 - r^2 + \alpha r$$

and

(4.2.11)
$$u_\varphi^2 = \frac{2\bar{a}\,(k^2-1)}{k^2 \alpha \bar{b}^2} + \frac{2\bar{a}}{k^2 \bar{b}^2} u - u^2 + \alpha u^3$$

and

(4.2.12)
$$u_{\varphi\varphi} + u = \frac{\bar{a}}{k^2 \bar{b}^2} + \frac{3}{2}\alpha u^2$$

If you choose for constant k the value

(4.2.13)
$$k^2 = \frac{\bar{a}}{\bar{a} + \frac{\alpha}{2}}$$

equations 4.2.10, 4.2.11 and 4.2.12 are identical with equations 2.8.71, 2.8.76 and 2.8.77 in reference /1/. Now we have derived for integration constants h and k values

(4.2.14)
$$h^2 = \frac{GM\bar{b}^2}{\bar{a} + \frac{\alpha}{2}}$$

(4.2.15)
$$k^2 = \frac{\bar{a}}{\bar{a} + \frac{\alpha}{2}}$$

With these values the results of reference /1/ and the results of Lawden book are identical. Insert expressions 2.1.14 and 2.1.15 into equations 4.2.14 and 4.2.15 and you get

(4.2.16)
$$h^2 = 2GM \frac{r_1^2 r_2^2}{(r_1 + r_2 + \alpha)(r_1 - \alpha)(r_2 - \alpha) - \alpha^3}$$

$$(4.2.17) \qquad k^2 = \frac{(r_1+r_2)(r_1-\alpha)(r_2-\alpha)}{(r_1+r_2)(r_1-\alpha)(r_2-\alpha)+\alpha\big((r_1-\alpha)(r_2-\alpha)-\alpha^2\big)}$$

Above the integration constants h and k are given as functions of known quantities that is as functions of perihelion and aphelion distances from the Sun and Schwarzschild radius. Take the classical limit (speed of light $c \rightarrow \infty$ so that $\alpha \rightarrow 0$) and you get

$$(4.2.18) \qquad h^2 = 2GM\frac{r_1 r_2}{r_1+r_2} = GM\frac{b^2}{a}$$

$$(4.2.19) \qquad k^2 = 1.$$

REFERENCES

/1/ U. Kivi (2020), Gravitation, Exact calculation of Newton and Einstein theory, 2. edited edition, Bod – Books on Demand, Helsinki, Finland

/2/ D.F.Lawden (1982), Introduction to Tensor Calculus, Relativity and Cosmology, Third Edition, Power Publications, Inc., Mineola, New York

APPENDIX 1: FORTRAN-PROGRAM

```
! Orbit.f90
!
! FUNCTIONS:
! Orbit – Entry point of console application.
!

!*******************************************************************************
!
! PROGRAM: Orbit
!
! PURPOSE:  Entry point for the console application.
!
!*******************************************************************************

    program Orbit

    implicit none

    ! Variables

    ! Body of Orbit

        !the 4-degree polynomi that is referenced in many places is the polynomi
        !under the square root in the integral form of the orbit equation
        !the polynomi is: R*(2*A*R*R-(B*B+R*R)*(R-ALF)), and the referenced
        !3-degree polynomi is 2*A*R*R-(B*B+R*R)*(R-ALF)

        !speed of light, mass of sun, gravitation constant, constant pi
        REAL(16) C, M, G, APII
        ! distance from sun, parameter, third root of 4-degree polynimi,
        classical distance (ellipse)
```

```
REAL(16) R, RA, R3, RCOS
!periheli, apheli, half 'major axis', square of half 'minor axis
REAL(16) R1, R2, A, B2
!original values for periheli, apheli, half 'major axis', square of half 'minor axis
REAL(16) RAA1, RAA2, AAA, BAA2
!Schwarzild, periheli shift, differential periheli shift, calculaion parameter,
calculaion parameter
REAL(16) ALF, EDI, DEDI, NJ, NJA
!polar angle inside a loop, classic polar angle inside a loop, total polar angle,
total classic polar angle
REAL(16) FII, FIICL, TFII, TFIICL
!3-degree polynomi argument and value at minimum and argument
and value at maximum
REAL(16) MIR, MIF, MAR, MAF
!zero points of 4-degree polunomi (numerical calculation)
REAL(16) S0, S1, S2, S3
!zero points of 4-degree polunomi (analytic calculation)
REAL(16) SB0,SB1,SB2,SB3
REAL(16) DN, DNA, DR, OSA(100), OSAS, QQ, QQQ, DPF, PF, PFCL
INTEGER(8) NA, NB, HROUNDS, MOD
INTEGER(8) K, N, I, I1, IA, IB, ITOT
INTEGER(8) NJAKAJA, IJAKAJA, IJAKAJAT(100), IJAKAJAI(100)

OPEN(UNIT=1,FILE='TULOS.TXT', DECIMAL = 'COMMA') !output to
Excel, planet orbit
OPEN(UNIT=2,FILE='KULMA.TXT') !output file
APII = 3.1415926535897932Q0 ! value of pi
G = 6.67428Q-11 ! gravitation constant
C = 2.99792458Q8 ! speed of light
M = 1.9891Q30 ! mass of Sun
!   M = 1.0Q6*M ! 1E6 mass of Sun
!   M = 1Q6*M ! 1E6X mass of Sun
!     M = 2.5220904Q6*M ! 2.5220904E6X mass of Sun
!     M = 3.0Q6*M ! 3E6 mass of Sun
```

```
!    M = 30.0Q6*M ! 3E6 mass of Sun
!    M = 90.0Q6*M ! 3E6 mass of Sun
!    M = 300.0Q6*M ! 3E6 mass of Sun
   R1 = 46001272Q3 ! planet periheli distance
   R2 = 69817079Q3 ! planet apheli distance
!    R1 = 16237038957.02175Q0 ! planet periheli distance
!    R2 = 90801333579.53153Q0 ! planet apheli distance
   MOD = 1 ! MOD = 1 = change parameters A and B, MOD = -1 = change
   periheli R1 and apheli R2
   HROUNDS = 24 !number of half rounds (= 180 degree)
!    QQ = direction – positive or negative – of radius R change in numerical integration
   QQ = -1.0Q0 !begin calculation from apheli
!   QQ = 1.0Q0 !begin calculation from periheli
   QQQ = QQ
   !integration nodalization for polar angle TFII calculation
   !integration argumant is the distance from sun = R
   NJAKAJA = 5 !nodal disivision divider
   NA = 1 !first major node in numerical integration
   NB = 7 !last major node
   OSA(1) = 1Q0 !relative lenght of first major node
   IJAKAJAT(1) = 1 !number of outputs in first major node
   IJAKAJAI(1) = 10000000 !number of subnodes between outputs in first major node
   OSA(2) = 1Q4
   IJAKAJAT(2) = 2
   IJAKAJAI(2) = 10000000
   OSA(3) = 1Q7
   IJAKAJAT(3) = 2
   IJAKAJAI(3) = 10000000
   OSA(4) = 1Q12
   IJAKAJAT(4) = 10
   IJAKAJAI(4) = 1000000
   OSA(5) = 1Q7
   IJAKAJAT(5) = 2
   IJAKAJAI(5) = 10000000
```

```
OSA(6) = 1Q4
IJAKAJAT(6) = 2
IJAKAJAI(6) = 10000000
OSA(7) = 1Q0 !relative lenght of last major node
IJAKAJAT(7) = 1 !number of outputs in last major node
IJAKAJAI(7) = 10000000 !number of subnodes between outputs in last major node
OSAS = 0
DO N= NA, NB
   IJAKAJAI(N) = NJAKAJA*IJAKAJAI(N)
   OSAS = OSAS + OSA(N)
END DO


ALF = 2Q0*M*G/(C*C) !Schwarzild radius
!    ALF = 0Q0 !calculation is made according to classical mechanics
A = (R1+R2)/2 !half of 'major axis'
B2 = R1*R2 !square of half of 'minor axis' = B**2


! let R1 and R2 be roots of the 4-degree polynomi then
! R3 is the smallest root of the 4-degree polynomi = about the Schwarzild radius
R3 = R1*R2*ALF/((R1-ALF)*(R2-ALF)-ALF*ALF)


CALL MINMAX !calculates the minimum and maximum of the four
degree polynomi
CALL POLYGRAPF !outputs thr graph of the four degree polynomi
!Calculate numerically the roots of the 4-degree polynomi
!R*(2*A*R*R-(B*B+R*R)*(R-ALF)) = 0
CALL NUMROOTS
!calculate the roots analytically
CALL ANAROOTS
CALL POLYGRAPF !outputs thr graph of the four degree polynomi
RAA1 = R1 !original value for periheli
RAA2 = R2 !original value for apheli
AAA = A !original value for half 'major axis'
BAA2 = B2 !original value for half 'minor axis' square B2 = B**2
```

```
! MOD = -1: change apheli and periheli to the largest roots of the 4-degree polynomi
IF(MOD.LT.0) THEN
    R1 = S2
    IF(MIF.GT.0) THEN !only two roots – S0 and S3
        R1 = S0
    END IF
    R2 = S3
! MOD = 1: change values of  A and B2 so that relativistic solution has the same
! periheli and apheli as the classical solution with the original A and B2
ELSE
    A = ((R1+R2)/2Q0)*(R1-ALF)*(R2-ALF)/((R1-ALF)*(R2-ALF)-ALF**2)
    B2 = R1*R2*R1*R2/(R1*R2-ALF*(R1+R2))
END IF


WRITE(2,*) 'Mercury orbit'
WRITE(2,*) 'mass of sun, gravitation constant, speed of light'
WRITE(2,1) M,G,C
WRITE(2,*) 'original periheli, apheli, parameter A = classically half major axis,
parameter B = classically half minor axis'
WRITE(2,1) RAA1, RAA2, AAA, QSQRT(BAA2)
WRITE(2,*) 'changed periheli, apheli, parameter A = classically half major axis,
parameter B = classically half minor axis'
WRITE(2,1) R1, R2, A, QSQRT(B2)
WRITE(2,*) 'Schwarzild'
WRITE(2,1) ALF
WRITE(2,*) 'minimi'
WRITE(2,1) MIR, MIF
WRITE(2,*) 'maksimi'
WRITE(2,1) MAR, MAF
WRITE(2,*) 'numerical roots'
WRITE(2,1) S0,S1,S2,S3
WRITE(2,*) 'analytical roots'
WRITE(2,1) SB0,SB1,SB2,SB3
```

```
WRITE(2,*) 'numeric integration nodal division'
WRITE(2,*) 'relative lenght of interwall, number of sub interwalls,
number of outputs'
DO I = NA, NB
   WRITE(2,1) OSA(I), IJAKAJAT(I)*IJAKAJAI(I), IJAKAJAT(I)
END DO
WRITE(2,*)
WRITE(2,*) 'calculation result'
WRITE(2,*) 'angle, rel radius , classic ellipse radius, difference (rel – classic)'
WRITE(2,*) 'classic angle (at the same radius as relativistic solution), periheli shift'

!     GOTO 10

!     RA = initial value for radius R
!     R1<R<R2
     IF(QQ.LT.0) THEN
         RA = R2 !begin the calculation from apheli
     ELSE
         RA = R1 !begin the calculation from periheli
     END IF
     TFII = 0 ! polar angle
     TFIICL = 0 ! classical polar angle (with same R)
     FII = 0 ! polar angle in a loop
     FIICL = 0 ! classical polar angle in a loop
     DN = R2-R1
     DPF = 0

     !do a separate calculation for the first time step to get output as
     !close to polar angle coordinate zero as possible
     IJAKAJA = IJAKAJAT(1)*IJAKAJAI(1)
     IF(IJAKAJA.EQ.0) THEN
         GOTO 10
     END IF
     DNA = DN*OSA(1)/OSAS
```

```
DR = DNA/IJAKAJA
R = RA+QQQ*DR/2Q0
FII =  FII+DR/LSA(R)
FIICL =  FIICL+DR/LSACL(R)
IF(QQ.LT.0) THEN
  RCOS = R1*R2/((R1+R2)/2+(R2-R1)*QCOS(FII+APII)/2) ! ellipse in polar
  coordinates
ELSE
  RCOS = R1*R2/((R1+R2)/2+(R2-R1)*QCOS(FII)/2) ! ellipse in polar coordinates
END IF
!   EDI = (FII-FIICL)*180/APII ! periheli shift in degrees
EDI = ((FII-FIICL)/(FIICL))*360*3600*100/0.2408Q0 !periheli shift in arc seconds
WRITE(*,1) FII*180/APII, R+DR/2, RCOS, R+DR/2-RCOS
WRITE(2,1) FII*180/APII, R+DR/2, RCOS, R+DR/2-RCOS
WRITE(*,1) FIICL*180/APII, EDI
WRITE(2,1) FIICL*180/APII, EDI

!do the calculation – numerical integration – for number HROUNDS half rounds
!that is for polar angle 0 < TFII < HROUNS*180 degree
DO K = 1, HROUNDS
DO N = NA, NB !main nodalization for R
  IJAKAJA = IJAKAJAT(N)*IJAKAJAI(N)
  IF(IJAKAJA.EQ.0) THEN
    GOTO 10
  END IF
  DNA = DN*OSA(N)/OSAS
  DR = DNA/IJAKAJA !delta R in numerical integration
  ITOT = 0 !counter
  FII = 0
  FIICL = 0
  DO I = 1, IJAKAJAT(N)
  FII = 0
  FIICL = 0
  DO I1 = 1, IJAKAJAI(N)
```

```
ITOT = ITOT + 1
R = RA+QQQ*(DR/2Q0+(ITOT-1)*DR) ! distance from sun
FII =  FII+1Q0/LSA(R) !polar angle coordinate
FIICL =  FIICL+1Q0/LSACL(R) !classical polar angle coordinate
DPF = DPF+DR/LSA(R)
!output relativistic orbit, classical orbit, and Schwarzild circle as x,y -coordinates
!and polar angle and periheli shift
IF(DPF.GE.APII/180) THEN
   PF = TFII+FII*DR
   PFCL = TFIICL+FIICL*DR
!        EDI = (PF-PFCL)*180/APII !perihel shift – degrees
EDI = ((PF-PFCL)/(PFCL))*360*3600*100/0.2408Q0 !perihel shift – arc seconds
   DEDI = (LSACL(R)/LSA(R)-1Q0)*360*3600*100/0.2408Q0 !differential
   perihel shift – arc seconds
   IF(QQ.LT.0) THEN
     PF = PF + APII
     PFCL = PFCL + APII
   ELSE
     PF = PF
     PFCL = PFCL
   END IF
   WRITE(1,1) R*QCOS(PF), R*QSIN(PF), R*QCOS(PFCL), R*QSIN(PFCL),
   ALF*QCOS(PF), ALF*QSIN(PF), PF*180/APII, EDI, DEDI
   DPF = 0
 END IF
END DO
TFII = TFII + FII*DR !polar angle coordinate
TFIICL = TFIICL + FIICL*DR !classical polar angle coordinate
IF(QQ.LT.0) THEN
 RCOS = R1*R2/((R1+R2)/2+(R2-R1)*QCOS(TFII+APII)/2) !classical
 distance from sun = ellipse
ELSE
 RCOS = R1*R2/((R1+R2)/2+(R2-R1)*QCOS(TFII)/2) !classical distance
 from sun = ellipse
```

```
      END IF
!     EDI = (TFII-TFIICL)*180/APII !perihel shift – degrees
      EDI = ((TFII-TFIICL)/(TFIICL))*360*3600*100/0.2408Q0 !perihel shift –
      arc seconds
      DEDI = (LSACL(R)/LSA(R)-1Q0)*360*3600*100/0.2408Q0 !differential
      perihel shift – arc seconds
      IF(QQ.LT.0) THEN
         PF = TFII + APII
         PFCL = TFIICL + APII
      ELSE
         PF = TFII
         PFCL = TFIICL
      END IF
      WRITE(1,1) R*QCOS(PF), R*QSIN(PF), R*QCOS(PFCL), R*QSIN(PFCL),
      ALF*QCOS(PF), ALF*QSIN(PF), PF*180/APII, EDI, DEDI
      WRITE(*,1) TFII*180/APII, R+QQQ*DR/2, RCOS, R+QQQ*DR/2-RCOS
      WRITE(2,1) TFII*180/APII, R+QQQ*DR/2, RCOS, R+QQQ*DR/2-RCOS
      WRITE(*,1) TFIICL*180/APII, EDI, DEDI
      WRITE(2,1) TFIICL*180/APII, EDI, DEDI
      END DO
      RA = RA+QQQ*OSA(N)*DN/OSAS
    END DO
    QQQ = -QQQ
!     EDI = (TFII-K*APII)*180/APII !perihel shift – degrees
    EDI = ((TFII-K*APII)/(K*APII))*360*3600*100/0.2408Q0 !perihel shift – arc seconds
    WRITE(2,*) 'periheli shift '
    WRITE(2,1) EDI
    WRITE(*,*) 'periheli shift'
    WRITE(*,1) EDI
    END DO
10    WRITE (*,*) 'END'

    CLOSE(UNIT=1) !output to Excel, planet orbit
    CLOSE(UNIT=2) !output file
```

```fortran
1     FORMAT(20(1X,G23.16))


      CONTAINS

      SUBROUTINE NUMROOTS
      !calculates numerically roots of polynomi R*(2*A*R*R-(B*B+R*R)*(R-ALF))
      REAL(16) NJA, X, Y, Z

      S0 = 0
      S1 = 0
      S2 = 0
      S3 = 0

      IF(MIF.GT.0) THEN
         GOTO 20
      END IF
      X = S0
      Y = MIR
        Z = (Y+X)/2Q0
        DO I = 1, 200
           NJA = Z*(2*A*Z*Z-(B2+Z*Z)*(Z-ALF))
           IF (NJA.LT.0) THEN
              Y = Z
           ELSE
              X = Z
           END IF
           Z =(X+Y)/2Q0
           WRITE(*,1) X,Y,Z,NJA
        END DO
      S1 = Z
```

```fortran
      X = MAR
      Y = MIR
        Z = (Y+X)/2Q0
        DO I = 1, 200
          NJA = Z*(2*A*Z*Z-(B2+Z*Z)*(Z-ALF))
          IF (NJA.LT.0) THEN
            Y = Z
          ELSE
            X = Z
          END IF
          Z =(X+Y)/2Q0
          WRITE(*,1) X,Y,Z,NJA
        END DO
      S2 = Z


20    CONTINUE

      X = MAR
      Y = 10*X
        Z = (Y+X)/2Q0
        DO I = 1, 200
          NJA = Z*(2*A*Z*Z-(B2+Z*Z)*(Z-ALF))
          IF (NJA.LT.0) THEN
            Y = Z
          ELSE
            X = Z
          END IF
          Z =(X+Y)/2Q0
          WRITE(*,1) X,Y,Z,NJA
        END DO
      S3 = Z

1   FORMAT(20(1X,G23.16))
```

END SUBROUTINE NUMROOTS

SUBROUTINE ANAROOTS
!calculates analytically roots of polynomi R*(2*A*R*R-(B*B+R*R)*(R-ALF))
REAL(16) A0,A1,A2,Q,APU
COMPLEX(16) T,S,D,R

A0 = -B2*ALF
A1 = B2
A2 = -(2*A+ALF)
Q = (3*A1-A2*A2)/9
R = (9*A2*A1-27*A0-2*A2*A2*A2)/54
D = Q*Q*Q+R*R
S = (R+SQRT(D))**(1.0Q0/3.00Q0)
T = (R-SQRT(D))**(1.0Q0/3.00Q0)
SB3 = -A2/3+REAL(S+T)
SB0 = 0
APU = 2*A+ALF-SB3
SB1 = (APU-QSQRT(APU**2-4*B2*ALF/SB3))/2Q0
SB2 = (APU+QSQRT(APU**2-4*B2*ALF/SB3))/2Q0

1   FORMAT(20(1X,G23.16))
END SUBROUTINE ANAROOTS

SUBROUTINE MINMAX
REAL(16) RR,RRQ

RR = (2*A+ALF)/3
RRQ = QSQRT(RR*RR-B2/3)
MIR = RR-RRQ ! R at polynomi minimum
MIF = 2*A*MIR*MIR-(B2+MIR*MIR)*(MIR-ALF) !value of polynomi at minimum
MAR = RR+RRQ ! R at polynomi maximum

MAF = 2*A*MAR*MAR-(B2+MAR*MAR)*(MAR-ALF) !value of polynomi at maximum

1 FORMAT(20(1X,G23.16))
END SUBROUTINE MINMAX

SUBROUTINE POLYGRAPF
!calculates the polynomi R*(2*A*R*R-(B*B+R*R)*(R-ALF)) graph
REAL(16) X, Y, Z, DX
OPEN(UNIT=3,FILE='POLYN.TXT', DECIMAL = 'COMMA') !output to Excel, four degree polynomi

DX = 2*MIR/200
X = -MIR
DO I = 1,200
   Y = X*((2*A*X*X-(B2+X*X)*(X-ALF)))/B2
   WRITE(3,1) X,Y,S0,S1,S2,S3
   X = X+DX
END DO
DX = (MAR-MIR)/200
X = MIR
DO I = 1,200
   Y = X*((2*A*X*X-(B2+X*X)*(X-ALF)))/B2
   WRITE(3,1) X,Y,S0,S1,S2,S3
   X = X+DX
END DO
DX = (S3-MAR)/200
X = MAR
DO I = 1,200
   Y = X*((2*A*X*X-(B2+X*X)*(X-ALF)))/B2
   WRITE(3,1) X,Y,S0,S1,S2,S3
   X = X+DX
END DO
DX = 0.1*S3/200

```fortran
      X = S3
      DO I = 1,200
         Y = X*((2*A*X*X-(B2+X*X)*(X-ALF)))/B2
         WRITE(3,1) X, Y, S0,S1,S2,S3
         X = X+DX
         IF(Y.LT.0) THEN
            GOTO 30
         END IF
      END DO
30  CONTINUE
      CLOSE(UNIT=3) !output to Excel, four degree polynomi

1   FORMAT(20(1X,G23.16))
      END SUBROUTINE POLYGRAPF


      REAL(16) FUNCTION LSA(R)
         REAL(16) R
         LSA = QSQRT((R*(2*A*R*R-(B2+R*R)*(R-ALF)))/B2)
      END FUNCTION LSA


      REAL(16) FUNCTION LSACL(R)
         REAL(16) R
         LSACL = QSQRT((R*((R1+R2)*R*R-(R1*R2+R*R)*R))/R1/R2)
      END FUNCTION LSACL


      end program Orbit
```

APPENDIX 2: MERCURY ORBIT

Mercury orbit

mass of sun, gravitation constant, speed of light

  0.1989100000000000E+31   0.6674280000000000E-10   299792458.0000000

original periheli, apheli, parameter A = classically half major axis, parameter

B = classically half minor axis

  46001272000.00000   69817079000.00000   57909175500.00000   56667637009.39376

changed periheli, apheli, parameter A = classically half major axis, parameter

B = classically half minor axis

  46001272000.00000   69817079000.00000   57909175500.00016   56667640028.17253

Schwarzild

  2954.266225571882

minimi

  18115285919.05607   -0.2611780232806819E+32

maksimi

  59096950050.45475   0.8296484452722052E+31

numerical roots

  0.000000000000000   2954.266540307006   46001257633.15288   69817093366.84681

analytical roots

  0.000000000000000   2954.266540307006   46001257633.15288   69817093366.84681

numeric integration nodal division

relative lenght of interwall, number of sub interwalls, number of outputs

1.000000000000000	50000000	1
10000.00000000000	100000000	2
10000000.00000000	100000000	2
1000000000000.000	50000000	10
10000000.00000000	100000000	2
10000.00000000000	100000000	2
1.000000000000000	50000000	1

calculation result

angle, rel radius , classic ellipse radius, difference (rel – classic)

classic angle (at the same radius as relativistic solution), periheli shift (integral and differential)

0.9301487270623908E-08	69817079000.00000	69817079000.00000	0.2381532247710627E-09
0.9301486578359050E-08	40.05607955179374		
0.9301089420033844E-04	69817078999.97618	69817078999.97619	-0.2033709337098964E-05
0.9301088727799505E-04	40.05602696590680	40.05602695827757	
0.6577798436475332E-02	69817078880.89953	69817078880.89966	-0.1263452966972289E-03
0.6577797946922147E-02	40.05602696485620	40.05602697769819	
0.9301948380927196E-02	69817078761.82288	69817078761.82305	-0.1682877129290899E-03
0.9301947688628929E-02	40.05602697129826	40.05602699711918	
0.2081957712174155	69816959685.17180	69816959685.15860	0.1319989352388401E-01
0.2081957557224258	40.05603344490493	40.05604641814883	
0.2942867233304303	69816840608.52071	69816840608.49167	0.2903872761832680E-01
0.2942867014281098	40.05603991861123	40.05606583924493	
30.28152007766668	67435307586.81657	67435307253.66863	333.1479419379187
30.28151781634843	40.19134737712191	40.45820584351760	
44.18108541112967	65053774565.11243	65053773936.18354	628.9288851469012
44.18108209967241	40.33957559021501	40.88978952237961	
55.97216949778092	62672241543.40828	62672240659.51668	883.8916016401676
55.97216528556431	40.50299212459906	41.35417339443588	
67.07013797203571	60290708521.70414	60290707426.85989	1094.844250910989
67.07013290200814	40.68456466176600	41.85524436436979	
78.13358461171985	57909175500.00000	57909174242.82749	1257.172506053502
78.13357867579197	40.88833428552439	42.39752873758780	
89.66331442790608	55527642478.29586	55527641114.70682	1363.589033419627
89.66330757743208	41.12012120287357	42.98632933697089	
102.2267150278731	53146109456.59172	53146108055.18496	1401.406759858858
102.2267071664480	41.38904720738968	43.62789948570179	
116.7346527648361	50764576434.88757	50764575089.86049	1345.027084807346
116.7346437177730	41.71155468559234	44.32966596272955	
135.3469470487371	48383043413.18343	48383042287.02727	1126.156163038082
135.3469364548062	42.12667909460285	45.10051780475199	
179.5533705887139	46001510391.47929	46001510376.20453	15.27475653371635
179.5533562445156	42.99631848283822	45.95118481718574	
179.6840317765371	46001391314.82820	46001391304.01357	10.81462998183142

179.6840174211832	42.99846719600077	45.95122964210671	
179.9858965825391	46001272238.17712	46001272237.69330	0.4838207104138415
179.9858822014124	43.00341949440436	45.95127437778905	
179.9900310810973	46001272119.10047	46001272118.75840	0.3420663901455884
179.9900166996176	43.00348720872011	45.95127442252530	
179.9998731769515	46001272000.02382	46001272000.01922	0.4594834324734613E-02
179.9998587946315	43.00364838885838	45.95127446726110	
180.0000143415050	46001272000.00000	46001272000.00025	-0.2457862967914908E-03
179.9999999591729	43.00365070052663	45.95132205831432	

periheli shift
42.88157627846215

180.0001555060584	46001272000.02382	46001272000.02890	-0.5082340545257474E-02
180.0001411237143	43.00365301219125	45.95127446726135	
180.0099976019126	46001272119.10047	46001272119.44278	-0.3423104905736969
180.0099832187282	43.00381417419882	45.95127442252530	
180.0141321004708	46001272238.17712	46001272238.66110	-0.4839861433032916
180.0141177169334	43.00388187038384	45.95127437778950	
180.3159969064728	46001391314.82820	46001391325.67495	-10.84674905811947
180.3159824971626	43.00881603808922	45.95122964210760	
180.4466580942960	46001510391.47929	46001510406.82306	-15.34377068986961
180.4466436738301	43.01094662062725	45.95118490665686	
224.6530816342728	48383043413.18343	48383045340.76380	-1927.580369713946
224.6530634635396	43.53200058540990	45.10051796651100	
243.2653759181738	50764576434.88757	50764579361.82186	-2926.934290269849
243.2653562005728	43.62368295792767	44.32966610966728	
257.7733136551368	53146109456.59172	53146113178.89383	-3722.302115243382
257.7732927518978	43.64396373187584	43.62789961976571	
270.3367142551038	55527642478.29586	55527646837.61630	-4359.320440160733
270.3366923409137	43.62836606275448	42.98632945978164	
281.8664440712901	57909175500.00000	57909180334.25436	-4834.254356221632
281.8664212425538	43.59001796203717	42.39752885050500	
292.9298907109742	60290708521.70414	60290713640.67039	-5118.966251173677
292.9298670163376	43.53463584866174	41.85524446854253	
304.0278591852290	62672241543.40828	62672246701.61995	-5158.211669821472

304.0278346327815	43.46402721606390	41.35417349084195	
315.8189432718802	65053774565.11243	65053779410.61420	-4845.501771180770
315.8189178186734	43.37633806827941	40.88978961185629	
329.7185086053432	67435307586.81657	67435311509.90319	-3923.086623960808
329.7184821019973	43.26193410838944	40.45820592678599	
359.7057419596795	69816840608.52071	69816840654.95938	-46.43866677507361
359.7057132169177	43.00606223479543	40.05606587808738	
359.7918329117925	69816959685.17180	69816959718.03211	-32.86031527671328
359.7918041626234	43.00535635750662	40.05604641814922	
359.9907267346290	69817078761.82288	69817078763.28965	-1.466764024649976
359.9906979706572	43.00372686538836	40.05602699711938	
359.9934508845734	69817078880.89953	69817078881.93609	-1.036554581091203
359.9934221203989	43.00370455949228	40.05602697769819	
359.9999356721157	69817078999.97618	69817078999.98861	-0.1242463015727563E-01
359.9999069074585	43.00365146208544	40.05602695827741	
360.0000286830099	69817079000.00000	69817078999.99774	0.2264651221956433E-02
359.9999999183458	43.00365070052663	40.05607955179374	

periheli shift
42.88157627846215

.

.

.

3960.000408524003	69817078999.97618	69817078999.54060	0.4355804885745141
3960.000092112691	43.00365063129403	40.05602695827757	
3960.006893311545	69817078880.89953	69817078869.20002	11.69951820635744
3960.006576899751	43.00364580435462	40.05602697769819	
3960.009617461490	69817078761.82288	69817078745.39156	16.43131877214705
3960.009301049492	43.00364377664347	40.05602699711918	
3960.208511284326	69816959685.17180	69816959323.25100	361.9207957396988
3960.208194857526	43.00349573862263	40.05604641814883	
3960.294602236439	69816840608.52071	69816840097.04799	511.4727232232440
3960.294285803232	43.00343166562884	40.05606583924493	
3990.281835590776	67435307586.81657	67435260434.88302	47151.93354779262
3990.281516918152	42.98230864412495	40.45820584351760	

4004.181400924239	65053774565.11243	65053713717.32100	60847.79142634089
4004.181081201476	42.97425599564654	40.88978952237961	
4015.972485010890	62672241543.40828	62672174196.32278	67347.08550236198
4015.972164387368	42.96879804980963	41.35417339443588	
4027.070453485145	60290708521.70414	60290639074.94204	69446.76209957816
4027.070132003812	42.96502673726492	41.85524436436979	
4038.133900124829	57909175500.00000	57909107237.17429	68262.82571132042
4038.133577777596	42.96272158355900	42.39752873758780	
4049.663629941015	55527642478.29586	55527578162.77935	64315.51650625528
4049.663306679236	42.96194760772860	42.98632933697089	
4062.227030540982	53146109456.59172	53146051694.48921	57762.10250659771
4062.226706268252	42.96301889673583	43.62789948570179	
4076.734968277945	50764576434.88757	50764528098.40391	48336.48366756908
4076.734642819577	42.96665237192112	44.32966596272955	
4095.347262561846	48383043413.18343	48383008696.05297	34717.13046068764
4095.346935556610	42.97466770541354	45.10051780475199	
4139.553686101823	46001510391.47929	46001510039.53051	351.9487807881690
4139.553355346319	43.00333266517797	45.95118481718574	
4139.684347289646	46001391314.82820	46001391065.86818	248.9600242471155
4139.684016522987	43.00342570921961	45.95122964210671	
4139.986212095648	46001272238.17712	46001272227.17720	10.99992093810348
4139.985881303216	43.00364064884069	45.95127437778905	
4139.990346594206	46001272119.10047	46001272111.36003	7.740436914425393
4139.990015801421	43.00364359256432	45.95127442252530	
4140.000188690061	46001272000.02382	46001272000.04255	-0.1873140683312691E-01
4139.999857896435	43.00365060001939	45.95127446726110	
4140.000329854614	46001272000.00000	46001272000.13002	-0.1300209510409283
4139.999999060976	43.00365070052663	45.95132205831432	

periheli shift
42.88157627846215

4140.000471019167	46001272000.02382	46001272000.26512	-0.2413064288756879
4140.000140225518	43.00365080103386	45.95127446726135	
4140.010313115022	46001272119.10047	46001272127.10070	-8.000231344433779
4140.009982320532	43.00365780845466	45.95127442252530	

4140.014447613580	46001272238.17712	46001272249.43675	-11.25963670066519
4140.014116818737	43.00366075214402	45.95127437778950	
4140.316312419582	46001391314.82820	46001391564.07990	-249.2516937450469
4140.315981598966	43.00387565749167	45.95122964210760	
4140.446973607405	46001510391.47929	46001510743.75663	-352.2773454579067
4140.446642775634	43.00396866726024	45.95118490665686	
4184.653397147382	48383043413.18343	48383078931.99323	-35518.80980287857
4184.653062565343	43.03201515952598	45.10051796651100	
4203.265691431283	50764576434.88757	50764626353.51555	-49918.62797347340
4203.265355302376	43.03953527007022	44.32966610966728	
4217.773629168246	53146109456.59172	53146169539.79335	-60083.20163773241
4217.773291853701	43.04278405053423	43.62789961976571	
4230.337029768213	55527642478.29586	55527709789.69726	-67311.40140489324
4230.336691442717	43.04357269657504	42.98632945978164	
4241.866759584399	57909175500.00000	57909247339.99214	-71839.99214082768
4241.866420344357	43.04261403061715	42.39752885050500	
4252.930206224083	60290708521.70414	60290781992.58361	-73470.87946837602
4252.929866118141	43.04022346653202	41.85524446854253	
4264.028174698338	62672241543.40828	62672313164.69805	-71621.28976048797
4264.027833734585	43.03647583220782	41.35417349084195	
4275.819258784989	65053774565.11243	65053839629.22611	-65064.11368640523
4275.818916920477	43.03117799376690	40.88978961185629	
4289.718824118452	67435307586.81657	67435358328.27805	-50741.46148107910
4289.718481203801	43.02350300868294	40.45820592678599	
4319.706057472788	69816840608.52071	69816841165.80522	-557.2845034807132
4319.705712318721	43.00385151111695	40.05606587808738	
4319.792148424902	69816959685.17180	69816960079.34185	-394.1700532762438
4319.791803264427	43.00379276325328	40.05604641814922	
4319.991042247738	69817078761.82288	69817078779.12327	-17.30038318188813
4319.990697072461	43.00365704744811	40.05602699711938	
4319.993766397682	69817078880.89953	69817078893.03786	-12.13833122008434
4319.993421222202	43.00365518869859	40.05602697769819	
4320.000251185225	69817078999.97618	69817078999.82632	0.1498607702339544
4319.999906009262	43.00365076398985	40.05602695827741	

4320.000344196119   69817079000.00000   69817078999.67389   0.3261097760051746
4319.999999020149   43.00365070052663   40.05607955179374
periheli shift
42.88157627846215

APPENDIX 3: SOLVING THE ROOTS OF THE 4-DEGREE POLYNOMIAL

Numerical solution

Root seek is done with the most simple method. Choose a segment of line on the r-axis that includes the zero point of polynomial f(r). At one of the end points of the line segment the value of the polynomial f(r) is positive and at the other one the value of the polynomial is negative. Bisect the segment of line and repeat the bisection for that segment of line that now includes the zero point. Continue to bisect the segment of line until the zero point has been determined with a wanted accuracy. Because of the simplicity of the algorithm the result converges fast and definitely, if you have found such a segment of line that includes the zero point and only that zero point.

Analytic solution

We shall find the zero points of the expression under square root in equation 2.1.6

(A.3.1) $$f(r) = r\left(2\bar{a}r^2 - (\bar{b}^2 + r^2)(r - \alpha)\right) = 0.$$

One root has value zero r = 0 so you have to solve a third order equation

(A.3.2) $$r^3 - (2\bar{a} + \alpha)r^2 + \bar{b}^2 r - \bar{b}^2\alpha = 0.$$

According to the formula for solution of the third order equation

(A.3.3) $$r^3 + a_2 r^2 + a_1 r + a_0 = 0.$$

one root is

(A.3.4) $$r_1 = -\frac{a_2}{3} + S + T$$

where

(A.3.5) $$S = \sqrt{R + \sqrt{D}}, \; T = \sqrt{R - \sqrt{D}}$$

(A.3.6) $$D = Q^3 + R^2, \; R = \frac{9a_2a_1 - 27a_0 - 2a_2^3}{54}, \; Q = \frac{3a_1 - a_2^2}{9}$$

In equation A.3.3

(A.3.7) $$a_0 = -\bar{b}^2\alpha, \; a_1 = \bar{b}^2, \; a_2 = -(2\bar{a} + \alpha)$$

so you can calculate values for Q and R

(A.3.8) $$Q = \frac{3\bar{b}^2 - (2\bar{a} + \alpha)^2}{9}$$

(A.3.9) $$R = \frac{18\bar{b}^2(\alpha - \bar{a}) + 2(2\bar{a} + \alpha)^3}{54}$$

and so you are able to calculate root r_1 using equation A.3.4.

If one knows one root r_1 of equation A.3.3 the other two roots may be calculated using the identity
(A.3.10)
$$r^3 - (2\bar{a} + \alpha)r^2 + \bar{b}^2 r - \bar{b}^2\alpha \equiv r^3 - (r_1 + r_2 + r_3)r^2 + (r_1r_2 + r_1r_3 + r_2r_3)r - r_1r_2r_3$$

and the result is

(A.3.11) $$r_2 = \frac{2\bar{a} + \alpha - r_1 + \sqrt{(2\bar{a} + \alpha - r_1)^2 - \frac{4\bar{b}^2\alpha}{r_1}}}{2}$$

(A.3.12) $$r_3 = \frac{2\bar{a} + \alpha - r_1 - \sqrt{(2\bar{a} + \alpha - r_1)^2 - \frac{4\bar{b}^2\alpha}{r_1}}}{2}$$

This solution method is used in the FORTRAN program to verify the numerical solution.